Laboratory and Skills Manual

EARTH SCIENCE

The Physical Setting

THOMAS McGUIRE
Earth Science Educator

AMSCO SCHOOL PUBLICATIONS, INC.,
a division of Perfection Learning®

Please visit our Web sites at:
www.amscopub.com and ***www.perfectionlearning.com***

The publisher wishes to thank the following educators who acted as reviewers for this project.

Deena Bollinger
Earth Science Teacher
South Orangetown Middle School
Blauvelt, NY

Diana K. Harding
Retired
New York State Education Department

Glenn Dolphin
Earth Science Teacher
Union–Endicott High School
Endicott, NY

JulieAnn Hugick
Earth Science Teacher
Eastchester Middle School
Eastchester, NY

Cover Design: Rosemarie Lanaro

Text Design: Mel Haber

Composition: Northeastern Graphic, Inc.

Art: Hadel Studio

Photo Credits:
FIGURE 9-1. Courtesy C.E. Meyer, U.S. Geological Survey
FIGURE 9-2. Courtesy U.S. Geological Survey
FIGURE 22-1. Courtesy NASA: The Remote Sensing Tutorial
FIGURE 23-1. Courtesy NOAA
All other photographs: Thomas McGuire

When ordering the Lab Book, please specify:
Either **14822** *or* **LABORATORY AND SKILLS MANUAL: EARTH SCIENCE—THE PHYSICAL SETTING**
ISBN 978-1-56765-910-8

Printed in the United States of America

9 10 11 12 13 PP 23 22 21 20 19

To the Teacher

Amsco's *Laboratory and Skills Manual for Earth Science* conforms to the New York State Core Curriculum for The Physical Setting/Earth Science, which is in line with the National Science Standards. Teachers can access the *Earth Science Reference Tables* at: http://www.nysdregents.org/testing/reftable/reftable.html

This book contains traditional laboratory exercises as well as inquiry-based Skill Sheets and Labs. In the Tips for Teachers, there are suggestions for making Skill Sheets and Labs more inquiry-based. Teachers can use these tips to modify their lessons to include more student-centered inquiry. The activities are designed to arouse students' curiosity and interest. The Tips for Teachers pages contain information to make your role more effective.

Clearly, the best way for students to learn science is for them to *do* science. This idea is central to an inquiry-based approach to science. While lectures and other teacher-led modes of instruction certainly have a place in science courses, participatory activities should be stressed. For example, you may ask students to design their own labs or ask volunteers to read aloud for the class the laboratory introductions or skill sheets text. Please keep in mind that factual content may be essential for assessment, but your influence on attitudes and values will probably be more important in helping your students to become life-long learners and informed citizens.

Each activity begins with the following information:

- **Level of Difficulty** will tell you how difficult students are likely to find learning the intended skills or performing the activity. The level assigned assumes that the teacher reads and follows the Suggestion for the Teacher. In some cases, that may include guiding students through the more challenging phases of the activity.
- **Content** refers to alignment with the New York State Core Curriculum and the Mathematics, Science, and Technology Standards.
- **Level of Interest** is an indication of how likely it is that students will find an activity or skill interesting. Providing engaging, challenging, and yet ultimately satisfying experiences for students is important to successful teaching and learning.
- **Preparations** offers suggestions and material needs to be considered before you do each activity. Suggestions may include demonstrations or motivations to introduce a topic and instructions on how to provide materials. Securing materials and equipment generally requires advanced preparation.

- **Materials** offers a list of materials teachers will likely require to conduct the Lab or Skill Sheet. This may include useful demonstrations.
- **Time** is an estimate of how much class time students will need to complete the Lab or Skill Sheet. These time estimates are based on the pacing used by the author. Adjust your own estimates accordingly. Many activities will take an entire class period, but others may be short enough to share a period with another activity or lesson. Some will take more than a single class period, but most can be completed on successive days.
- **Suggestions for the Teacher** presents hints to make an activity more successful for students and teachers. For example, you might project an important diagram onto a chalkboard to discuss it with the class and even to write notes on it. These tips may help you avoid potential pitfalls.
- **Extensions** offers ideas on how to make an activity more inquiry-based and suggestions for further research. Some of these can be assigned for homework.

We hope you and your students find these activities relevant, engaging, challenging, and ultimately satisfying.

Contents

Science Laboratory Safety Contract

As a participant in a laboratory science course, I understand that lab safety is essential. As my teacher provides safety rules and instructions for laboratory exercises, I will follow these instructions. In addition, I will . . .

- never perform any experiment or action unless it can be done safely.
- protect myself, especially my eyes, face, and hands, by wearing appropriate safety equipment.
- not harm others while conducting laboratory procedures.
- follow housekeeping practices in the laboratory and science room.
- know the location of fire-fighting and safety equipment.
- know where to get help in case of an emergency.
- conduct myself in a responsible manner at all times in class and laboratory situations.
- never perform any procedures unless authorized by my teacher.
- report all accidents or unsafe conditions to my teacher immediately.

I, _____, have read and agree to follow the safety regulations above. I further agree to follow all other written and verbal instructions that are needed to ensure the safety of myself and others.

Date: _____ Signature: _____

Parental Acknowledgment: _____ Date: _____

Adapted from the National Science Teachers Association Safety in the Secondary Science Classroom.

CHAPTER 1—SKILL SHEET 1: PERCENT DEVIATION

Use the following steps to solve percent deviation problems:

1. Write the algebraic formula. You may copy it exactly from the *Reference Tables* or you may abbreviate some terms as shown below. Remember that "show your work" means that you *must* start with the algebraic formula.

2. Substitute the values of the measurements. It is best to use the numbers and the units.

3. Show your numerical answer as a percent (%):

$$Deviation\ (\%) = \frac{difference\ from\ accepted\ value}{accepted\ value} \times 100$$

4. Express your answer to the appropriate number of significant figures.

5. The terms "percent deviation" and "percent error" are often used interchangeably.

Sample Problem

A student measured the length of a table as 1.9 meters. If the length of the table was really 2.0 meters, what was the percent deviation of the student's answer?

Solution

$$Dev\ (\%) = \frac{dev}{acc\ val} \times 100$$

$$= \frac{2.0\ m - 1.9\ m}{2.0\ m} \times 100$$

$$= \frac{0.1\ m}{2.0\ m} \times 100$$

$$= 0.05 \times 100$$

$$= 5.0\%$$

Practice Problems

(Solve the following problems in the space provided.)

1. The volume of a 10-mL solid was estimated to be 8 mL by a student. What was the percent deviation of the student's answer?

2. A stone measured as 25 kg was actually 20 kg. What was the percent deviation of this measurement?

3. When asked to try to hold your hands 100 cm apart, you actually held them 104 cm apart. What was your percent deviation in this challenge?

4. I tried to estimate the length of one minute. My friend timed me and reported that my guess was really 66 seconds. What was my percent deviation?

5. A reference book lists the population of a town as 1995 people. A careful count found 2000 residents. What is the percent deviation of the published figure?

6. A student calculated the density of a sample of quartz as 2.97 g/cm³. If the published density of quartz is 2.70 g/cm³, what was the percent deviation of the answer?

7. A tree that is actually 10 meters tall was estimated to be 25 meters tall. What was the percent deviation?

8. A student asked to pick out a 1-kg stone selected one with a mass of 950 grams. What was the percent error?

9. If I were to guess your age to be 12 years, what would be my percent deviation?

10. Which is a larger percent error, 1 cm in a meter, or 1 g in a kilogram? (Please justify your answer by showing the calculations.)

CHAPTER 1—SKILL SHEET 2: EXPONENTIAL NOTATION

In science we often use very large or very small numbers. For example, the average distance between Earth and the nearest star outside our solar system is 40,000,000,000,000 miles in standard notation. The diameter of the nucleus of an atom is about 0.000 000 01 centimeter in standard notation. Powers of 10, sometimes called exponential notation, or scientific notation, provides a convenient way to express and work with these numbers. You probably know that $10^2 = 100$ and $10^3 = 1000$. Similarly, $10^1 = 10$ and $10^0 = 1$. (Any number to the power "0" is equal to 1.) In addition, $10^{-1} = \frac{1}{10}$ and $10^{-3} = \frac{1}{1000}$.

Any number can be expressed in terms of powers of 10. For example:

$$7 = 7 \times 1 = 7 \times 10^0$$
$$53 = 5.3 \times 10 = 5.3 \times 10^1$$
$$26,900 = 2.69 \times 10^4$$

Express the following numbers as standard numbers:

1. $10^2 =$ _____   **2.** $10^4 =$ _____   **3.** $10^0 =$ _____   **4.** $10^{-3} =$ _____

Every number written in exponential notation has three parts:

$$\text{coefficient} \longrightarrow 3.61 \times 10^8 \quad \begin{matrix} \nearrow \text{exponent} \\ \searrow \text{base} \end{matrix}$$

The coefficient is always a positive or negative number whose value is 1 or greater but less than 10. The coefficient contains all significant digits in the original number. The base will always be 10 in the exponential notation we will be using. The exponent is equal to the number of places that the decimal has been moved to convert the number into the coefficient. Numbers greater than 1 will have positive exponents. A negative power of 10 does not mean a negative number. A negative exponent just means a number less than 1 but greater than 0. For example:

$$3 \times 10^{-4} \text{ is equal to } 0.0003$$

To write a negative number in exponential notation, the negative sign must be placed in front of the coefficient.

$$-7 \times 10^2 = -700$$

Write the following numbers in exponential notation:

5. $1000 =$ _____   **6.** $\frac{1}{1000} =$ _____

7. $35 =$ _____   **8.** $-50 =$ _____

9. $5500 =$ _____

10. $1{,}000{,}000 =$ _____

11. $0.08 =$ _____

12. $-0.003 =$ _____

13. Write the following number in correct exponential notation form:

745×10^4. _____

Exponential notation makes it easier to read and to write very large or small numbers. Exponential notation is especially useful with the metric system. The following table provides the meanings of some of the prefixes used in the metric system. The three that are in **_bold italic_** type are the most commonly used prefixes. Use this information complete the list that follows the table.

Prefix	Meaning	Numerical Multiplier
giga-	10^9	one billion
mega-	10^6	one million
kilo	**_10^3_**	**_one thousand_**
centi-	**_10^{-2}_**	**_one-hundredth_**
milli-	**_10^{-3}_**	**_one-thousandth_**
micro-	10^{-6}	one-millionth
nano-	10^{-9}	one-billionth

1 meter $= 100$ centimeters $= 10^2$ centimeters

1 meter $=$ _____ millimeters $=$ _____ millimeters

1 centimeter $=$ _____ millimeters $=$ _____ millimeters

1 centimeter $= \frac{1}{100}$ meter $= 10^{-2}$ meter

1 millimeter $=$ _____ meter $=$ _____ meter

1 meter $=$ _____ kilometer $=$ _____ kilometer

Remember that:

1. The value of the coefficient is always 1 or greater, but less than 10.

2. The exponent is the number of places that the decimal point was moved to make the coefficient.

3. Negative exponents are used for small numbers: those between 0 and 1.

CHAPTER 1—LAB 1: DENSITY OF GRANITE

$$Density = \frac{mass}{volume}$$

Introduction

The density of a material determines how heavy it feels. Dense substances have a large amount of matter packed into a relatively small volume. We can therefore define density as the concentration of matter.

Density is a mathematical relationship. It is the ratio of mass to volume. Density can be calculated by dividing the mass of an object by its volume. Density also determines the ability of a substance to sink or float in water. Dense materials, such as lead and rock, sink. Materials of low density, such as air and balsa wood, float. Any object that is less dense than the fluid (liquid or gas) in which it is resting will float. Any object that is more dense than the fluid in which it is resting will sink.

Solid objects made of the same material usually have the same density. If the object is cut into pieces, or if a larger or a smaller sample is selected, the density will remain constant.

Objective

To determine the density of several pieces of granite, and compare your results with the accepted value.

Materials

3 pieces of granite of different sizes

Procedure

1. Obtain a single piece of granite. (It can be any size.)

2. Measure mass and volume of your piece of granite. Use these values to calculate its density. (Use the formula already given.)

3. Return the first piece of granite. Repeat the procedure with two different-sized samples of the granite.

Relative Size	Mass (grams)	Volume (cm³)	Density (g/cm³)
Small			
Medium			
Large			
		Total of Densities	
		Average Density	

The accepted value for the density of granite is 2.7 g/cm³. Calculate the percent deviation of your average density. Show your work here.

Wrap-Up

1. What are the most common metric units of density?

   _____

2. How does the size of the sample affect the density?

   _____

3. No matter how much granite you have, the density should always be approximately

   _____

4. Define density:

   _____

   _____

5. Why does ice float in water?

   _____

   _____

6. Of the six measurements you made, which do you think is the least accurate. Explain your choice. (Remember the percent deviation calculation above.)

   _____

   _____

CHAPTER 1—LAB 2: DENSITY OF FLUIDS

Introduction

 A fluid is any substance that can flow, which includes liquids and gases. This lab exercise will show you how to construct a graph to represent the density of different substances.

 Record all data and your lab report on a sheet of lined notebook paper. You will be graded on readability, neatness, and accuracy.

 Hints

 A. 1 mL = 1 cm³.

 B. Instead of trying to pour an exact number of milliliters of the fluid, concentrate on making your readings exact.

 C. When you measure the volume of the fluids, read the bottom of the meniscus.

Objective

 To determine the density of two fluids

Materials

 Fluids A and B, 100-mL graduated cylinder, mass scale

FIGURE 1-1. Mass scale

FIGURE 1-2. Graduated cylinder

Procedure

 1. On lined notebook paper, record the mass (Figure 1-1) of an empty 100-mL graduated cylinder (Figure 1-2).

 2. *a.* Obtain a bottle of fluid A or fluid B. (Please take just one at a time.)

  b. Pour approximately 20 to 30 mL of one of the fluids into a graduated cylinder. Carefully record the volume and determine the mass. (Remember to subtract the mass of the graduated cylinder.)

    c. Add about 10 to 15 mL, and measure the mass and volume again. Repeat this until you have measured a total of five different amounts.

    d. Repeat the procedure with the other fluid. (That is a total of five different amounts of fluid A and five different amounts of fluid B.)

3. Graph your results, with volume in mL (or cm³) on the horizontal axis and mass in grams on the vertical axis. (You must include the units.) Construct the two lines (one for fluid A and another for fluid B) as best-fit lines on a single graph. Best-fit lines show the trend of the data, but they may not quite connect the points. (Your teacher can help you make best-fit lines.)

Wrap-Up ▌▌▌➡

1. Which of these two fluids you measured is more dense?

_____

2. How does the graph show which of the fluids is more dense?

_____

_____

3. What is the density of fluid A? (Hint: It is water.)

_____

4. Which is more dense, 10 g of water or 100 g of water? Explain your answer.

_____

5. From your graph, accurately determine the mass of 10 mL of fluid B.

_____

6. What do you think was your greatest source of error during data collection?

_____

_____

How could the error be reduced?

_____

    Submit to your teacher the data table, the graph, and the answers to questions 1–6 in complete sentences.

Chapter 2
Earth's Dimensions and Navigation

CHAPTER 2—SKILL SHEET 1: CLUES TO EARTH'S SHAPE

Although there are still a few people who think that Earth is flat, most people know that our planet is a sphere. See Figure 2-1. However, as with other facts that you will learn this year, the most important issue is not what you know, but how you know it is true.

For thousands of years it was obvious that Earth must be flat. If it were not, people on the other side would fall off. When we began to realize how gravity pulls us toward the center of planet Earth, we could better understand a round planet. In fact, gravity has given us an important method to determine the exact shape of Earth.

Gravity becomes increasingly weaker the further you are from Earth's center. With sensitive instruments, scientists can measure tiny changes in gravity. For example, the pull of gravity is slightly weaker high in the mountains than it is at sea level. Over the whole Earth, the strength of gravity varies by about 1%. Gravity is slightly stronger at the poles than it is near the equator. There are two reasons for this. First, the spin of Earth creates a slight force, which can counteract gravity. Second, and more importantly, Earth bulges slightly at the equator. A person at the equator is about 20 km farther from Earth's center than a person at the poles. See Figure 2-2.

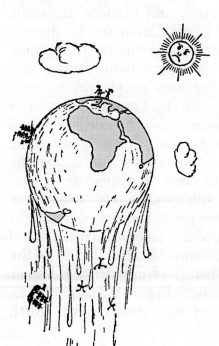

FIGURE 2-1. An argument against the round Earth.

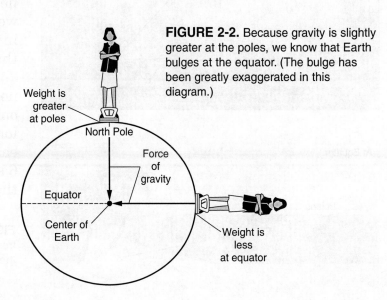

FIGURE 2-2. Because gravity is slightly greater at the poles, we know that Earth bulges at the equator. (The bulge has been greatly exaggerated in this diagram.)

Weight is greater at poles

North Pole

Force of gravity

Equator

Center of Earth

Weight is less at equator

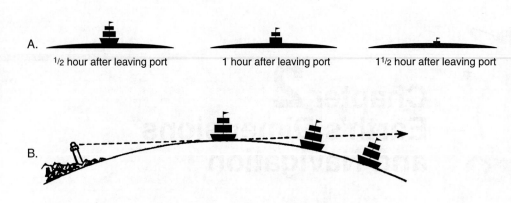

A.

1/2 hour after leaving port   1 hour after leaving port   1 1/2 hour after leaving port

B.

FIGURE 2-3. As a tall ship puts out to sea, it appears to sail "over the horizon." Viewed through a telescope, the ship would disappear from the bottom up, as shown in Part A. (The ships to the right are increasingly magnified.) As shown in Part B, Earth's curvature is responsible for this observation.

The true shape of Earth has been described as a spheroid, very slightly flattened at the poles and very slightly bulging at the equator. Please note that the flattening has been greatly exaggerated in Figure 2-2. In fact, Earth looks like a perfect sphere from every direction.

Long before the voyages of Columbus, people noticed that a distant ship seems to sail "over the horizon." On a clear day, and especially through a telescope, a ship can be seen to disappear from the bottom up, as it "sinks" below the horizon. See Figure 2-3. This effect can be seen when you look out to sea in any compass direction. It is caused by the curvature of Earth's surface. More than 2000 years ago, some enlightened people interpreted this to mean that Earth is a sphere.

A third way to determine the shape of Earth is to make observations of stars and planets. If people over the whole Earth try to observe the sun at any particular time, some will see it, and some (where it is night) will not. For some people, the sun will be high in the sky. For others it will be near or even below the horizon. When we account for all these observations in a logical model, we can determine that Earth must be a sphere.

On a spherical planet, when it is noon in one place, it must be midnight on the opposite side of Earth, 180° of longitude away. Meanwhile, 90° to the east or west, the time will be 6 A.M. or 6 P.M. Consider Figure 2-4. A person at the equator can observe the North

At North Pole

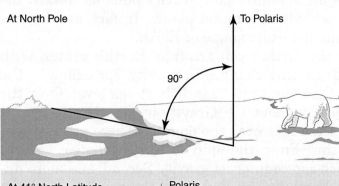

To Polaris

90°

At 41° North Latitude

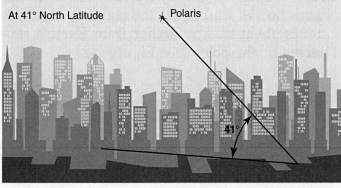

Polaris

41°

At Equator

Polaris

FIGURE 2-4. As an observer travels north from the equator, the North Star moves higher in the sky.

FIGURE 2-5. An eclipse of the moon, when the moon moves into Earth's shadow.

Star along the horizon. As she or he travels northward, Polaris will rise in the sky. Finally, at the North Pole, it will be directly overhead.

When the moon moves in its orbit into the shadow of Earth, we observe an eclipse of the moon. Because the moon is smaller than Earth, we cannot see the whole shadow. However, the edge of Earth's shadow is always round. See Figure 2-5. The sphere is the only shape with a shadow that is always round. In this way, observations of lunar eclipses support the spherical shape of Earth.

FIGURE 2-6. Earth looks perfectly spherical.

So far, we have considered only indirect evidence. The space age has allowed scientists to make direct observations and take photographs of our home planet. It was no surprise to observe the nearly spherical shape of planet Earth. See Figure 2-6. In fact, Earth is so close to a perfect sphere that we cannot detect the slightest flattening without the use of precision instruments.

1. State four ways to determine Earth's shape without leaving the ground.

   a. _____

   _____

   b. _____

   c. _____

   d. _____

2. How close to a perfect sphere is Earth's true shape?

   _____

3. At the equator, how high in the sky is the North Star?

   _____

4. Why do objects weigh slightly more at Earth's poles than they do at the equator?

   _____

   _____

My Notes

CHAPTER 2—SKILL SHEET 2: CELESTIAL NAVIGATION

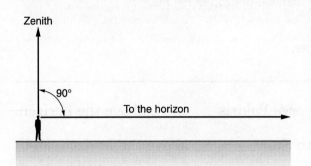

FIGURE 2-7. The zenith is the imaginary point in the sky straight overhead.

Before the invention of GPS technology, how were people on ships far at sea, out of the sight of land, able to tell where they were? For thousands of years mariners have navigated their ships by the stars. The procedure of finding your position through observations of the stars is called celestial navigation. Understanding how it works is not very difficult.

It is important that you work through this activity step by step. Proceed slowly and with care. If you do not understand something, review the earlier steps or ask for help.

The point in the sky straight overhead is known as the zenith. The zenith is at an angular elevation of 90° above the horizon. Figure 2-7 illustrates the meaning of zenith.

1. Define zenith. _____

2. The zenith is located _____ above the horizon.

3. In Figure 2-8, which observer will see Polaris at the zenith? _____

4. **a.** Does everyone, no matter where they are on Earth, see the same stars in the night sky?

   _____ (See Figure 2-8.)

   b. Explain

   _____

   _____

5. At the North Pole, an observer would see Polaris _____ above the horizon.

6. An observer at the equator would see Polaris _____ above the horizon, and at the equator.

7. **a.** In Figure 2-8, would man C be able to observe Polaris? _____

   b. Explain

   _____

8. Polaris is visible only to observers north of

   _____

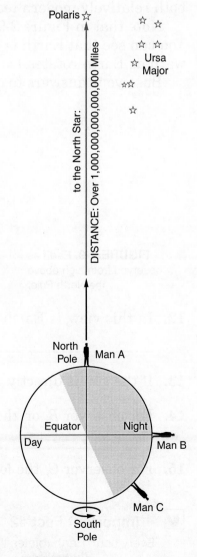

FIGURE 2-8.

Important Fact #1

Your latitude north of the equator is equal to the angle from the horizon to Polaris.

9. Copy the statement above in the space below.

_____

10. An observer at 40° North latitude could observe Polaris _____ above the horizon.

11. A person at the equator would see the North Star _____ above the horizon, and his or her latitude is _____.

Finding latitude is simple. Just measure the angle from the northern horizon up to Polaris. Polaris is easy to find because, for us, it is always in the same position in the northern sky. However, determining longitude is more difficult. Because our observations are made from a rotating earth, the sun and all the other stars appear to move through the sky. Determination of your longitude requires the use of a very accurate clock or a radio, both relatively modern technologies.

Note that in Figure 2-9 you are looking down from a point high above the North Pole. You can see that Earth is turning to the east. (That is, any location on Earth moves eastward as Earth rotates.)

Base your answers to questions 12–15 on Figure 2-9.

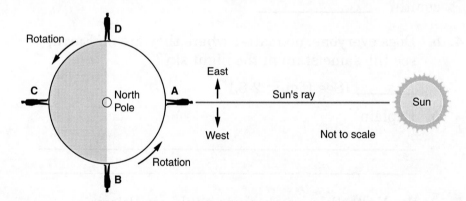

FIGURE 2-9. Earth as observed from high above the North Pole.

12. In this view, is Earth rotating clockwise or counterclockwise?

_____

13. If the sun is directly overhead for observer *A*, his local time must be _____.

14. For observer *B*, on the equator where the sun is just rising, the local time is _____.

15. For observer *C*, the local time is _____.

 Important Fact #2

Every hour, Earth rotates through 15° of longitude.

16. Please copy Fact #2 below.

_____

Because different places around Earth have different local times, you can determine your longitude. Follow the steps below, then, answer questions 17–20.

Step 1: On the outside of the circle in Figure 2-10, complete the labeling at 30° intervals from 30° to 360°. (Some have been done for you.)

Step 2: Notice that inside the circle some of the intervals have been labeled like a 24-hour clock.

Step 3: Complete labeling the inside of the circle in two-hour intervals. (Some have already been done for you.)

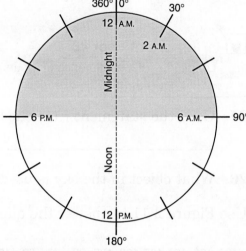

FIGURE 2-10.

17. Every hour Earth rotates through _____.

18. _____ means the same as "spin."

Earth can be split into time zones based on degrees of longitude.

19. Around the whole Earth there are _____ or _____ hours.

20. Thus, two places 15° of longitude apart are _____ apart in time.

If you look at very old maps, you will find that the measurements of latitude are accurate. However, distortion results from inaccurate determinations of longitude. The determination of longitude depends upon the use of a clock set to standard time. For this use, the ship's clock had to remain accurate after many weeks at sea. An error of only one hour would mean an error of 15° of longitude on a map.

The British dominated exploration and map-making in the 1600s and 1700s. They set their clocks to observations of the sun made at the Royal Observatory in Greenwich, near London, England. Greenwich Mean Time (GMT) therefore became the standard upon which longitude was based. Accurate maps could not be made until mariners had precise clocks (chronometers) that would keep accurate time on a long ocean voyage. If the navigator knew Greenwich time and his local time, based upon observations of the sun and stars, he could calculate his longitude.

21. To what time did the navigators set their clocks?

_____

If a ship's clock set to Greenwich Mean Time (GMT) reads 12 noon, but it takes one more hour for the sun to reach its highest point in the sky (local noon), the ship must be 1 hour, or 15°, west of the Prime Meridian.

22. If the local time is 3 hours behind GMT, the ship's longitude must be _____ West. (Hint: 3 × 15°)

23. If the local time is 2 hours ahead of GMT, the ship is _____ East.

24. If GMT and the local time are 12 hours apart, the ship is _____ from the Prime Meridian.

Longitude is calculated by comparing local time with Greenwich Mean Time.

 Important Fact #3

Longitude = time difference (in hours) from GMT × 15° per hour.

25. Copy the statement from the box above into the space below.

_____

26. What object in the sky is used to determine the local time? _____

Use Figure 2-11 to answer the questions 27–39.

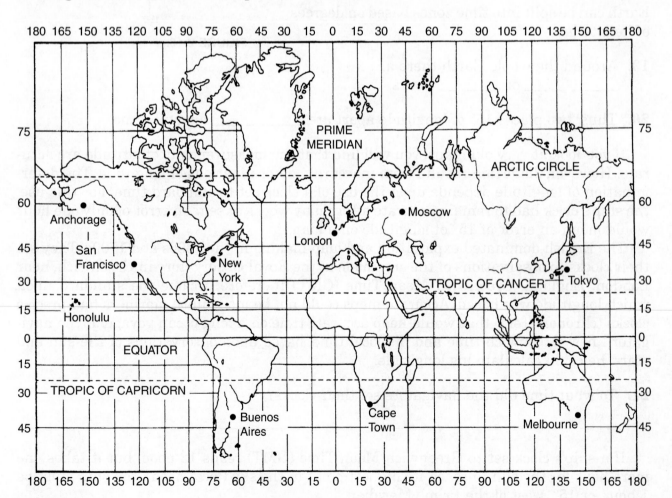

FIGURE 2-11.

Complete the table below.

Terrestrial Coordinates

Name of City	Latitude	Longitude
27. _____	35°N	140°E
28. _____	33°S	18°E
29. _____	55°N	37°E
30. London	_____	_____
31. Melbourne	_____	_____
32. Honolulu	_____	_____

33. During one day, the sun appears to move from east to west. Circle the name of the city at which noon will come first. *San Francisco   New York*

34. On the same day, is it noon in Moscow *before   or   after* at is noon in London. Circle the correct word.

35. A person in Rome, 42° North latitude, and a person in Chicago, also 42° North latitude, both see Polaris on the same night. Which, if either, will see it higher in the sky? Explain your answer

_____

36. What is the angular elevation of Polaris in your location?

_____

37. If it is noon in London, what time is it in New York? _____

38. If it is noon in Tokyo, what time is it in Melbourne? _____

39. How high above the horizon is Polaris in Cape Town, South Africa?

_____

40. Briefly explain a way to find your latitude using your own observations.

_____

41. Briefly explain a way to find your longitude by using your own observations.

_____

_____

42. Figure 2-12 contains information used by an observer to find her terrestrial coordinates.

    a. What is the observer's latitude? _____, longitude? _____

    b. Where is this person located? _____

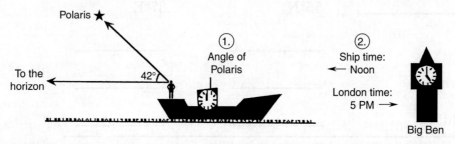

FIGURE 2-12.

43. As an observer travels due east or due west, what happens to the angle of Polaris?

_____

44. Lines of constant _____ run east and west, but they measure how far north or south you are.

45. The _____, is the reference line for latitude, and the _____, is the reference line for measurements of longitude. What are the terrestrial coordinates (latitude and longitude) at the point where they meet? _____

46. Define latitude:

_____

47. Define longitude:

_____

☀ CHAPTER 2—LAB 1: DETERMINING YOUR LATITUDE

Materials

Soda straw; cardboard, or file folder; coin, or small weight; masking tape and/or glue; string, scissors

Objective

To show your parent or guardian how to determine your latitude, without the use of any communication devices (like a computer) or printed materials (maps, reference books, etc.)

Procedure

Figures 2-13 and 2-14 will help you to construct and use an astrolabe. An astrolabe is a device to measure the angular altitude of any point in the sky. You may either make this device following the steps below, or you may just explain to the adult how it is made and used.

1. Cut out the protractor below and paste it to a sheet of thin cardboard.

2. Attach a coin or weight to a piece of string and hang it so that it pivots from the place labeled "String swings from here" at the center of the paper protractor.

3. Attach a drinking straw along the flat side of the protractor with tape as indicated in Figure 2-13.

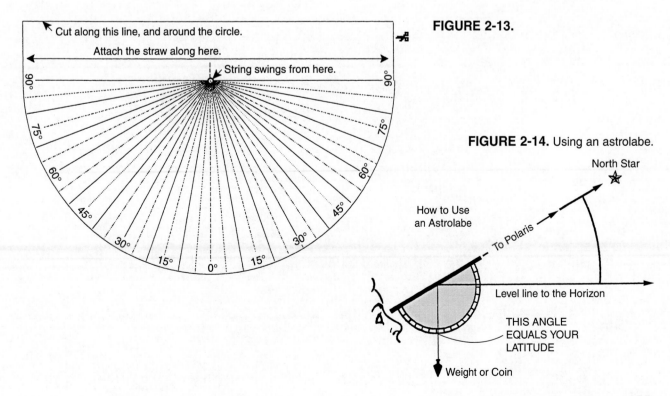

FIGURE 2-13.

FIGURE 2-14. Using an astrolabe.

4. Outside at night, use the "pointer stars" at the end of the Big Dipper to find Polaris. (You may need to find a location such as a park where at least a few stars are visible and not obscured by nearby lights or buildings.) Figure 2-15 shows how to use the pointer stars to find the North Star.

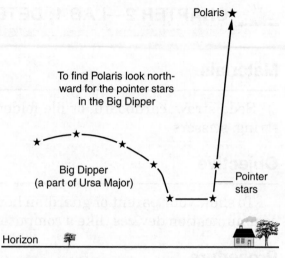

FIGURE 2-15. Using the Big Dipper to find Polaris.

5. Look through or along the straw pointed to Polaris as shown in Figure 2-14.

6. Ask another person to determine the latitude by reading the angle along the string.

The angle is _____.

My student, _____, has shown me how to determine our latitude.

(Signed) _____

Chapter 3
Models and Maps

 CHAPTER 3—SKILL SHEET 1: DRAWING ISOLINES

Isolines are lines that connect places that have the same field quantity value within a field region. The field value can be land elevation, temperature, strength of gravity, or any other measurement that can be made continuously over a defined area. Most isolines follow a few simple rules. Answer the questions below using the simple isoline map, Figure 3-1.

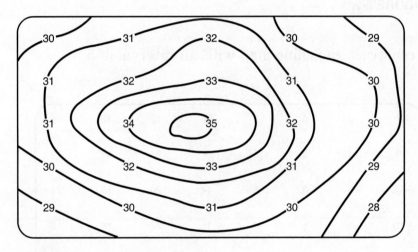

FIGURE 3-1. A simple isoline map.

1. Do the isolines ever touch or cross each other? _____

2. Do isolines usually have sharp angles or gentle curves ? (Please circle one.)

3. What does each point on any one isoline have in common with all other points on the same isoline?

_____

4. Do isolines ever end, except at the edge of the map? _____

5. Is the change in value from one isoline to an adjacent isoline always the same on a single map? _____

6. Do isolines tend to make parallel curves? _____

7. Does every isoline have one side where the values are higher and another side where the values are lower? _____

It is best to use a pencil whenever you draw isolines. Nearly everyone finds that they have to make erasures as a part of correcting errors. Draw isolines on Figure 3-2 at an interval of 1. Please work neatly.

13	15	16	16	18
12	13	14	15	17
11	12	13	14	15
10	12	13	15	16

FIGURE 3-2.

On Figure 3-3, construct an isoline map with an interval of 5.

−7	−2	0	2	8	12	16	9
−9	−4	3	5	9	15	15	10
−11	−5	5	8	14	20	18	11
−10	−4	5	11	15	22	18	10
−5	0	6	10	14	15	13	8
−3	0	7	9	10	8	4	2
−5	−1	5	5	5	2	−1	−6

FIGURE 3-3.

8. Define isoline:

_____

_____

CHAPTER 3—SKILL SHEET 2: READING TOPOGRAPHIC MAPS

A topographic, or contour, map is a two-dimensional model of the three-dimensional shape of the land surface. On a topographic map, isolines connect points of equal elevation. Contour maps also include a wide variety of geographic information. You should try to imagine the topography as you look at the contour lines on these maps.

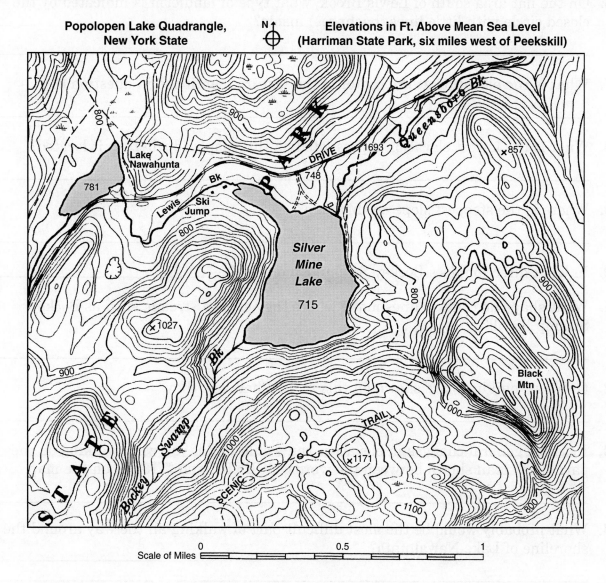

Popolopen Lake Quadrangle,
New York State

N ↑

Elevations in Ft. Above Mean Sea Level
(Harriman State Park, six miles west of Peekskill)

FIGURE 3-4.

Use the map segment in Figure 3-4 to answer the following questions.

1. What named geographic feature is in the southeastern part of the map?

2. What is the length of Lake Nawahunta?

3. What is the change in elevation between Silver Mine Lake and Lake Nawahunta?
   _____

4. What is the contour interval on this map? _____

5. What is the elevation of the top of Black Mountain? _____

6. On the flat area south of Lewis Brook, what type of landform is indicated by the closed circle with little lines (hachures) inside?

   _____

7. What is the elevation at the center of the feature mentioned in question 6?

   _____

8. What is the lowest elevation represented on this map and where is it located?

   _____

9. Where is the steepest slope on this map?

   _____

10. In what direction does Bockey Swamp Brook flow? _____

11. What shape do contour lines make where they cross a stream, such as Bockey Swamp Brook?

   _____

12. Only two buildings are shown on this map. Where are they?

   _____

13. Describe general changes in elevation you would find if you were walking on the part of the trail shown between the words "SCENIC" and "TRAIL" on the map.

   _____

14. What probably would be the most difficult part of walking all the way around the shoreline of Lake Nawahunta?

   _____

15. A friend has called you on her cell phone. She was hiking the Scenic Trail and injured herself rather badly. Fortunately, she has a map like the one above. She is located at the 800-foot contour on the trail. You have called an ambulance. They know how to get to Queensboro Brook but need detailed directions about how to get a stretcher and medical personnel to your friend. They do not have a topographic map. Write detailed directions to be read to the ambulance crew.

   _____

   _____

CHAPTER 3—LAB 1: A TOPOGRAPHIC MAP IN THREE DIMENSIONS

Introduction

A topographic map shows the shape of the land surface with contour lines. On page 29 you will find Figure 3-6, a contour map. You will be using this map to make a three-dimensional model of the land at this location. You will cut the map along the contour lines, paste each piece of the map on a sheet or cardboard or foam, and stack the pieces of cardboard or foam.

For this lab, the number of people in your group may be determined by the number of contour lines (lines of equal elevation) to be cut out in making your model of the land. The procedure below should make this clear.

Objective

The object of this lab is to help you visualize the shape of the land just by looking at a flat topographic map.

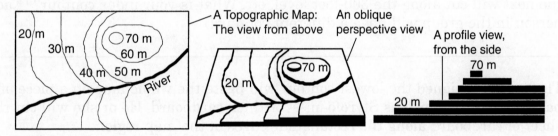

FIGURE 3-5. A contour model from three points of view.

Materials

Contour map on page 29, foam tray or corrugated cardboard, scissors, glue or glue sticks, highlighter pens or colored pencils

Procedure

1. What is the difference in elevation (contour interval) from one thin contour line to the next?

2. What is the minimum (lowest) elevation on the map on page 29? (Hint: It is less than 250 feet.)

3. What is the maximum elevation shown in the region of your map? (Hint: It is more than 550 feet.)

_____

4. What is the range in elevation? (This is the difference between the highest elevation and the lowest.)

_____

5. How many dark 50-foot index contour lines are shown on the map?_____

6. Using this interval, how many contour levels will you need to cut out? (*Hint:* the answer to step 5 plus 1.) _____ This is the maximum number of stacked sheets of foam or cardboard and the approximate number of people in your group for this activity.

7. Now that you have determined the number of people in your group, you will need to assign each person a different contour level, starting at the base level, which is about 200 feet on this map. The next person will cut along the 250-foot level, and the next will cut along the 300-foot level, etc. What is your index contour? (Each person in the group will have a different contour!)

_____

8. The person assigned the lowest contour will paste the whole map on a piece of stiff backing material such as Styrofoam™ or thick cardboard. He or she will cut the sheet of cardboard along the rectangular edges of the map region.

9. Cut out your map along your assigned contour, and then paste the paper to the backing material. Cut the backing along your assigned contour line to isolate the areas at the index level or above. (Note: Do not cut along the roads or the streams. Cut only along the contour line.)

10. When each person in the group has cut out her or his own piece of cardboard, paste the sections together starting with the lowest contour level and ending with the highest contour level to make a three-dimensional model of the map area.

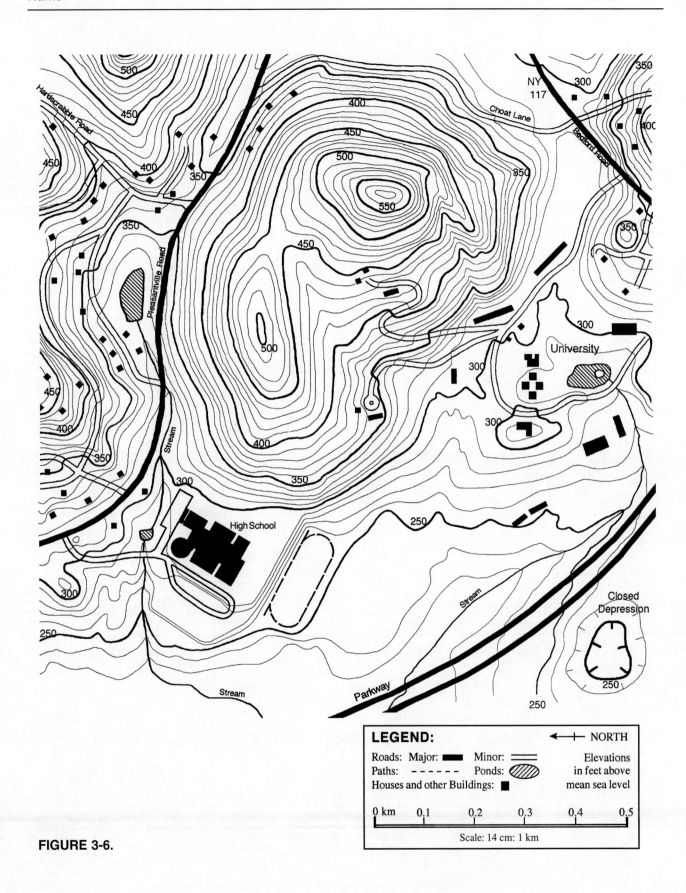

LEGEND: ← NORTH

Roads: Major: ▬ Minor: ═   Elevations
Paths: – – – – – Ponds: ⬭   in feet above
Houses and other Buildings: ■   mean sea level

0 km   0.1   0.2   0.3   0.4   0.5

Scale: 14 cm: 1 km

FIGURE 3-6.

Wrap-Up

(You will need your three-dimensional topographic model or the full map page to answer the following questions.)

1. This map is a model; of what is it a model?

_____

2. What name is applied to the lines on a topographic map that show the elevation of the land?

_____

3. What do contour maps show about the map area that is not shown on other maps, such as most road maps?

_____

_____

4. What does any point on a particular topographic contour line have in common with all other points on the same line?

_____

5. Explain what the contour interval on a topographic map is.

_____

6. *a.* What is the contour interval between the thin lines?

   b. What is the interval between the thicker index contour lines?

7. What is the elevation of the high school?

_____

8. What is the elevation of the pond along Pleasantville Road?

_____

8. Describe the shape of the land in places where the contour lines are far apart, such as the area near the oval running track south of the high school.

_____

10. How can you tell where the slope is steep just by looking at contour lines?

_____

11. How long is the portion of Bedford Road shown on this map?

_____

12. What is the minimum straight-line distance from Bedford Road to the nearest corner of the high school?

_____

13. Near the bottom right corner of the map are contour lines with hachures (little lines that point inward). What does this special kind of contour line show?

_____

14. What is the elevation of the land at the center of this feature?

_____

15. What shape is made by contour lines that cross a stream?

_____

16. In what direction, do the angular shapes mentioned in question 17 always point?

_____

17. If you wanted to walk from the pond on Pleasantville Road to the pond at the university with only a small change in elevation, what route would you take?

_____

_____

18. If an injured person were stranded at the top of the highest hill, what route could an ambulance take to get as close as possible to that person?

_____

19. If Pleasantville Road were completely blocked by an accident along the pond, how many different ways could you drive around the roadblock and still stay within this map?

_____

20. In what way is this map unlike most maps that you might use?

_____

_____

CHAPTER 3—SKILL SHEET 3: AN OCEAN BOTTOM PROFILE

A profile is a cutaway view, a cross section, or a silhouette. Topographic profiles show the shape of the land, with the highest places representing hills and the lowest places valleys. A profile shows the change in elevation along a particular path. In some profiles, the vertical scale has been enlarged or exaggerated to show changes in elevation more clearly. This greatly increases the slope. However, without vertical exaggeration, the changes in elevation would be difficult to visualize. Figure 3-7 is a cross section that has a vertical exaggeration of 200 times the horizontal distance.

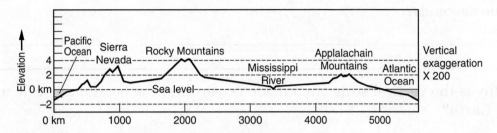

FIGURE 3-7. Profile of North America at about 41° North.

Use the data below to construct a profile of the Atlantic Ocean. You will need a sheet of graph paper. Start by labeling a horizontal line near the top "Sea Level." Measure all depths below sea level from this reference line. The largest numbers indicate the greatest depths. Label the left side of your ocean profile "North America" and right side "Europe." Also label the "Great Abyssal Plain" and the "Mid-Atlantic Ridge" at the proper positions on the profile paper. Complete your graph before you answer the following questions.

Distance (km)		Depth (km)	Distance (km)		Depth (km)
0	Cape Cod, MA	0	4000	The	3.0
200		0.2	4100	Mid-Atlantic	2.5
400		0.6	4200	Ridge	0.3
600		2.1	4300		1.1
900		3.0	4400		0.5
1200		3.5	4500		2.7
1500	The Great	3.4	4800		3.1
2000	Abyssal	3.6	5300		2.8
3000	Plain	3.1	5800	Land's End,	0.1
(Data continues above.)			6100	Great Britain	0

FIGURE 3-8. Ocean-depth data.

1. What two terms/expressions have the same meaning as profile?

_____

2. Your profile shows the total thickness of the (a) atmosphere, (b) hydrosphere, (c) lithosphere. (Circle your answer.)

3. The ocean bottom is (a) all flat, (b) totally mountainous, (c) a region that has both features (Circle your answer.)

4. A submerged mountain range runs down the center of the Atlantic Ocean. What is the name of this geographic feature?

_____

5. What landform would occur if this feature extended above sea level?

_____

6. According to the *Earth Science Reference Tables,* where does this occur?

_____

7. Why does the profile line of North America rise, while the Atlantic Ocean profile line descends?

_____

_____

8. Why is the vertical scale often made larger than the horizontal scale in profiles of Earth?

_____

9. Calculate the vertical exaggeration of your profile. (Please show your work.)

_____

_____

_____

_____

_____

CHAPTER 3—LAB 2: TOPOGRAPHIC PROFILES

Introduction

A profile is a cross-sectional view. A topographic profile shows the hills and valleys as high and low places on the profile.

Objective

To create a topographic profile from a topographic map

Materials

Pencil, blank edge of paper (this can be a narrow strip, as shown below, or as whole sheet of paper), Figure 3-9—Erinburgh regional topographic map

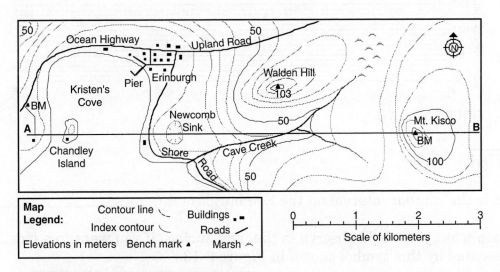

FIGURE 3-9. Erinburgh regional topographic map.

Procedure

1. Lay the blank edge along path A–B on the Erinburgh topographic map (Figure 3-9). This is illustrated in Figure 3-10.

2. Make a small mark along the edge of the paper at each place the edge crosses or touches a contour line. Label the marks with the contour line elevation as shown in Figure 3-10.

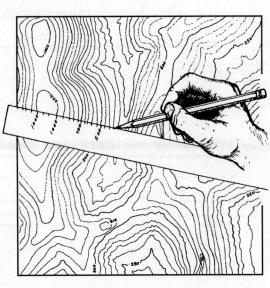

FIGURE 3-10. Marking the edge of the paper.

3. Figure 3-11 shows you how to use the marked edge of your paper to place a point at the indicated elevation exactly above each mark on the grid below (Figure 3-12).

4. Connect the dots on the grid to finish the profile line.

5. Label the following features along your A–B profile line on the grid: Chandley Island, Kristen's Cove, Newcomb Sink, Cave Creek, and Mt. Kisco.

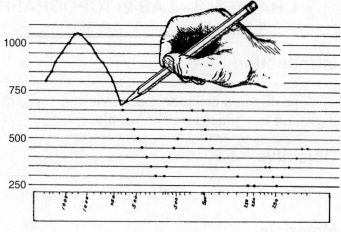

FIGURE 3-11. Marking points along the grid.

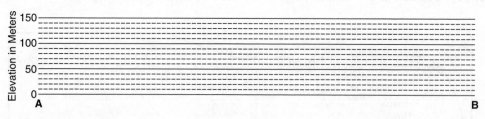

FIGURE 3-12. Grid.

Wrap-Up

1. What is the contour interval on the Erinburgh map? _____

2. Explain what you would observe in the land surface if you went to a location represented by this symbol shown in Figure 3-13.

_____

FIGURE 3-13.

3. What is the horizontal map scale in metric units?

_____

4. What is the vertical scale on the profile in metric units?

_____

5. What is the vertical exaggeration of the Erinburgh profile?

_____

_____

_____

CHAPTER 3—SKILL SHEET 4: CALCULATING GRADIENT

Gradient is a synonym for slope. If a hill has a large gradient, it is a steep hill, which changes quickly in elevation. On any isoline map, the areas where the isolines are the closest together indicate places with the steepest gradient.

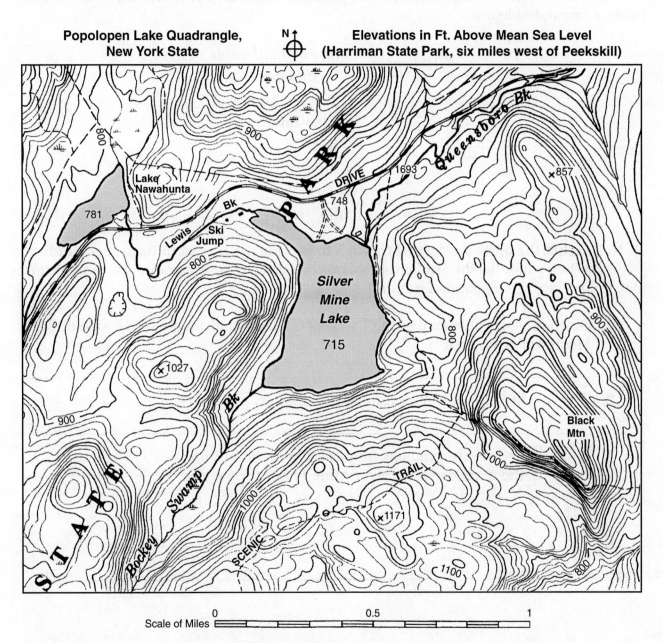

Popolopen Lake Quadrangle, New York State

N

Elevations in Ft. Above Mean Sea Level (Harriman State Park, six miles west of Peekskill)

Scale of Miles 0　　0.5　　1

FIGURE 3-14. Popolopen Lake Quadrangle, New York State. Elevations are in feet above mean sea level. (Harriman State Park, 6 miles west of Peekskill)

To calculate gradient, use this formula given in the *Earth Science Reference Tables:*

$$Gradient = \frac{change\ in\ field\ value}{distance}$$

For example, the average straight-line gradient from Lake Nawahunta to Silver Mine Lake is calculated below:

$$Gradient = \frac{change\ in\ field\ value}{distance}$$

$$= \frac{781\ feet - 715\cdot feet}{0.4\ mile}$$

$$= \frac{66\ feet}{0.4\ mile}$$

$$= \frac{165\ feet}{mile}$$

Notice that the units are carried through to provide the correct units of gradient in the answer. When you solve the gradient problems below show your work in this manner. (You may abbreviate the algebraic formula.)

1. On the topographic map of the Popolopen Lake Quadrangle, which side of Black Mountain is steepest?

    _____

2. Which side of Black Mountain has the smallest gradient?

    _____

3. What is the average gradient from the top of the hill marked 1171 (bottom center of the map) down to Silver Mine Lake? (Show your work.)

4. Calculate the average straight-line gradient of the portion of Bockey Swamp Brook shown on the map above. (Show your work.)

5. What is the average gradient from the top of Black Mountain straight north to the edge of the map? (Show your work.)

Gradient can be calculated for any field value. For example, if the atmospheric pressure in Chicago is 1007 millibars (mb), and the air pressure 500 km away at New York City is 993 mb, what is the pressure gradient between these two cities?

$$Gradient = \frac{change\ in\ field\ value}{distance}$$

$$= \frac{1007\ mb - 993\ mb}{500\ km}$$

$$= \frac{14\ mb}{500\ km}$$

$$= \frac{0.028\ mb}{km}$$

6. The difference in air pollution at two locations 1 km apart is 20 particles per cubic meter. What is the pollution gradient?

7. On a cold winter day, the temperature is 20°C at the center of a room, but only 18°C at a window that is 3 meters away. What is the temperature gradient?

8. A synonym for gradient is _____

9. What is the contour interval on the Popolopen Lake map in this activity?

_____

10. On any isoline map, how can you tell where the gradient is steepest?

_____

Chapter 4
Minerals

☼ CHAPTER 4—SKILL SHEET 1: MINERAL PROPERTIES

Nearly all rocks are composed of the elements, compounds, and mixtures that Earth scientists call minerals. A **mineral** is defined as a natural chemical solid of inorganic origin that has well-defined properties and a specific range of composition. That is, each mineral has a unique and uniform chemical make up, which gives it uniform chemical and physical properties. We can test for these properties to identify minerals. There are thousands of minerals, but most of them are rare. The great majority of the rocks that we see are composed of only about a dozen of the most common minerals.

FIGURE 4-1. Malachite is a copper ore mineral.

Some rocks contain no minerals. For example, coal is made of carbon from the accumulation of the remains of fossil plants. Because of its organic origin, coal contains no minerals. Some limestone is derived from the hard parts of shellfish and coral. This kind of limestone may therefore contain few or no minerals. On the other hand, ice is a mineral because it fits the definition above. Yet, no rocks contain ice as a mineral constituent.

Some rocks contain only a single mineral. Quartzite is composed of quartz, either pure or with minor impurities. Marble is predominantly calcite. However, most rocks contain a variety of minerals. Granite usually contains feldspar, quartz, mica, and amphibole. Other minerals, such as magnetite and pyroxene, may also be present.

1. What are nearly all rocks made of? _____

2. How many different minerals are there? _____

3. Of the thousands of minerals, how many are very common? _____

Minerals are identified by their observable properties. Geologists have selected certain observations that are most useful in identifying different minerals. **Color** is one of the most readily observed characteristics. Some minerals are easy to identify by their color.

Copyright © 2007 AMSCO School Publications, Inc.

Almandine (a variety of garnet) is always dark red. Pyrite is a brassy yellow. However, many light-colored minerals can be discolored by small amounts of impurities. For example, quartz may be colorless (clear), white, pink, green, brown, or even black. In addition, some different minerals are the same color; for example white quartz and calcite. Therefore, although color is easy to see, it can also be misleading.

4. Why is color of limited use in identifying minerals?

_____

5. What causes many light-colored minerals to display a wide variety of colors?

_____

6. Name one mineral that can be almost any color.

_____

FIGURE 4-2. Selenite gypsum has a glassy luster, while pyrite has a metallic luster.

Luster is the way that light behaves at the surface of a mineral. Geoscientists usually characterize luster as metallic (shiny with no entry of light) or nonmetallic. It is important to note that light does not penetrate the surface of a mineral that has a metallic luster. Minerals with this kind of luster look like they are made of a hard metal. Transparent or translucent surfaces cannot have a metallic luster. Nonmetallic lusters include glassy, pearly, waxy, and earthy (dull). It takes a little practice and a fresh mineral surface to correctly identify luster; yet, it can be one of the most useful properties in mineral identification. When you observe luster, ask yourself, "Does this look like it could be made from a hard metal?"

7. What is luster? _____

8. What is the luster of fresh aluminum foil? _____

9. What types of luster are shiny, but not metallic?

_____

Most mineral samples do not contain perfectly shaped crystals. However, when a mineral sample does contain crystals, **crystal shape** can be very useful in identifying minerals. Calcite and quartz are commonly transparent and colorless with a glassy luster. However, quartz forms hexagonal (six-sided) crystals, while calcite usually forms rhombohedral crystals that look like a cube pushed over toward one corner.

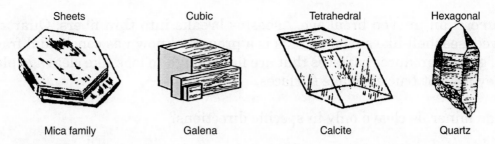

FIGURE 4-3. Each of these minerals shows a characteristic crystal shape.

Crystal shapes are determined by the arrangements and bonding of atoms and molecules. For example, halite (rock salt) crystals are cubic or rectangular solids. This is because the molecules in halite are locked into flat rows, layer upon layer. The molecules in calcite are in offset rows. Therefore, calcite crystals are not cubic with 90° corners, but rhombohedral.

10. What name is applied to the geometric and symmetrical shape of minerals?

_____

11. What do crystals of table salt look like? _____

12. What is the most common crystal shape of calcite _____

13. What determines the shape of the crystals of a particular mineral?

_____

Cleavage is the way a mineral splits, generally along flat planes. Cleavage depends on the arrangement and bonding of molecules. Minerals tend to split along the planes of weak bonds between their atoms. In specifying the cleavage properties of a mineral, scientists count the number of nonparallel planes of cleavage, and the angle between those cleavage planes. For example, the mica minerals split in one direction, forming thin sheets. Halite (rock salt) crystals cleave in three directions at right angles (90°) to each other. Many minerals cleave parallel to the crystal faces. However, some, like quartz, do not split parallel to any crystal face.

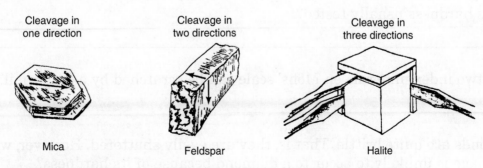

FIGURE 4-4. Number of cleavage directions is one way to specify cleavage.

Fracture is an uneven breakage. Asbestos breaks into thin fibers. Quartz fractures along curved, seashell-like surfaces. This is a property known as conchoidal fracture. The fracture of garnet produces surfaces that are flat enough to look like cleavage planes, even though they are not true cleavage surfaces.

14. Why do minerals cleave only in specific directions?

_____

15. What property is similar to cleavage, but the breakage is not controlled by the weak atomic bonds?

_____

16. What are the angles at which cleavage surfaces meet in halite? _____

The **hardness** of a mineral is determined by using it to scratch other solids. We test hardness by drawing the pointed edge of the unknown mineral across a clean surface of a known substance. A substance will scratch only materials that are either softer or have the same hardness. If the known substances is not scratched by the mineral, the mineral is softer. Figure 4-5 shows Mohs' Scale of Hardness. Hardness is determined by the strength of atomic and molecular bonding in a mineral. The diagrams in the figure illustrate the hardness of several common materials

MOHS' SCALE OF HARDNESS

1. Talc	6. Feldspar
2. Gypsum	7. Quartz
3. Calcite	8. Topaz
4. Fluorite	9. Corundum
5. Apatite	10. Diamond

Fingernail 2.5   Penny 3   Window glass 5.5   Steel file 6.5

FIGURE 4-5. Mohs' Scale of Hardness

17. What is the hardest mineral on Mohs' scale? _____

18. How is hardness usually tested?

_____

19. What two index minerals on Mohs' scale can be scratched by a fingernail?

_____

20. Diamonds are quite brittle. That is, they are easily shattered. However, what kind of damage is unlikely to occur to a diamond because of its hardness?

_____

The **streak** test shows the color of the powder of a mineral. We usually test the streak by rubbing a corner of the mineral across a white, unglazed porcelain streak plate. Minerals that have a metallic luster often leave a streak that is a different color from the surface of the sample. For example, pyrite is brassy yellow in color. However, the streak of pyrite is green to black.

Specific gravity is the ratio of the density of a mineral to the density of water. As a ratio, specific gravity has no units. However, specific gravity is the same number as the density in grams per cubic centimeter. Many common minerals have a specific gravity (density) in the range of 2.5–3 (a density of 2.5–3 g/cm³).

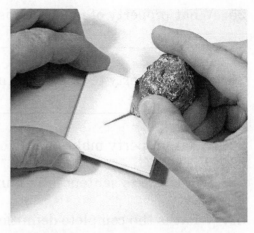

FIGURE 4-6. To test for streak, rub the sample across a porcelain streak plate.

21. What test allows us to observe the color of the powder of a mineral sample?

_____

22. Specific gravity compares the density of a mineral to the density of what other substance?

_____

23. For what group of minerals is the streak test especially useful?

_____

24. Why is there no unit of measure included in specific gravity?

_____

There are a number other properties that are found in only a few minerals. These special properties can make those minerals very easy to identify. For example, transparent crystals of calcite can break light into two images (**double refraction**) as shown in Figure 4-7. Uranium (uranite) is a mineral that shows **radioactivity**. (See Figure 4-8.)

Other minerals also have some unusual properties. Calcite **bubbles with acid** if a strong enough acid solution is applied. Magnetite is the only common mineral that is strongly attracted by a magnet. (It is strongly **magnetic**.) Halite (rock salt) has a **salty taste**. (In general, you should not taste substances in the science laboratory, unless instructed to do so.)

FIGURE 4-7. Transparent calcite shows double refraction.

FIGURE 4-8. Uranium ore is radioactive.

25. What property of magnetite is unusual, but easy to test?

_____

26. What property of calcite can make one line appear as two lines?

_____

27. What property makes halite easy to identify? _____

28. Words are to sentences as minerals are to _____

29. What is the complete definition of a mineral?

_____
_____

☀ CHAPTER 4—LAB 1: MINERAL IDENTIFICATION

Introduction

Define mineral.

Objective

To identify minerals and a rock by determining their properties and comparing them with a mineral identification chart.

Materials

Mineral identification set: 15 minerals, 1 rock, streak plate, glass bottle or plate, magnet on a string, mineral identification chart, hand lens

Procedure

1. Obtain a set of minerals from your teacher. Arrange the samples, in order, on the table. If the set is incomplete or if you have extra objects, please tell your teacher so the sets are maintained in good order.

2. Look at each mineral sample. Fill in the mineral identification grid on page 000 for each mineral. Make sure you list the name and characteristics next to the correct number on the grid. List only the characteristics that you observe. Ask your teacher whether you should perform destructive tests, such as cleavage, acid reaction, etc.; otherwise, do not destroy the samples.

DATA TABLE. MINERAL IDENTIFICATION

Sample Number	Mineral Name	Luster	Cleavage or Streak	Other Distinguishing Characteristics

3. Carefully examine sample 16 with a hand lens. This is a rock that contains a variety of minerals. Based on your observations of the mineral samples, list the four (4) most abundant minerals in rock sample number 16.

_____

_____

Wrap-Up ▶

The following minerals are specifically named in the New York State Earth Science Core Document. It is important that you be able to recognize the characteristics of each of these five minerals. You should do this with the mineral sample in front of you, so you can directly observe the most important characteristics.

1. Quartz: Color _____

   Luster _____

   Hardness _____

   Cleavage, Fracture, or Streak _____

Other Distinguishing Properties _____

2. Feldspar (Potassium and/or Plagioclase. They can be hard to distinguish.):

Color _____

Luster _____

Hardness _____

Cleavage, Fracture, or Streak _____

Other Distinguishing Properties _____

3. Calcite: Color _____

Luster _____

Hardness _____

Cleavage, Fracture, or Streak _____

Other Distinguishing Properties _____

4. Mica (Muscovite or Biotite):

Color _____

Luster _____

Hardness _____

Cleavage, Fracture, or Streak _____

Other Distinguishing Properties _____

5. Magnetite: Color _____

Luster _____

Hardness _____

Cleavage, Fracture, or Streak _____

Other Distinguishing Properties _____

6. Circle the characteristic that best distinguishes each of these minerals from other minerals.

If your mineral set is complete, including the 16 samples, streak plate, magnet, scratch glass, and hand lens, return it to your teacher. If it is not complete, please tell your teacher what is extra or missing.

7. Sample 16 obviously was a rock and not a single mineral. How could you tell that it is not one mineral?

_____

8. If there are thousands of minerals, why is it important to learn to identify only about a dozen?

_____

9. What determines the shape of the crystals of any particular mineral?

_____

10. Quartz is often used for jewelry and decoration, to make glass, and as the precision timekeeper in watches and clocks. Name two properties of quartz that you observed. For each property, name a different possible use for quartz that would depend on that property.

_____

_____

11. Name three minerals that you are you likely to come into contact with indoors, every day.

_____

_____

12. Identify each mineral crystal shown in Figure 4-9. Enter the names next to number in the chart below. *(The first one has been done for you.)*

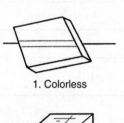

1. Colorless

2. Colorless

3. White

4. Pink to White

5. Black

6. Shiny gray

FIGURE 4-9.

#	Name
1	Calcite
2	
3	
4	
5	
6	

Chapter 5
The Formation of Rocks

 CHAPTER 5—LAB 1: IGNEOUS ROCK IDENTIFICATION

Introduction

All igneous rocks were formed by the solidification of molten magma or lava. Volcanic rocks are composed of very small crystals because the molten rock cooled quickly at or near Earth's surface. The rapid cooling does not give the crystals enough time to grow large. Volcanic rock is sometimes called extrusive rock because the molten rock (magma) is extruded, or pushed out, at Earth's surface, where it is known as lava. For this reason, all volcanic rocks have a fine-grained texture. Texture includes such characteristics as the grain size and shape, and any pattern such as layering. Layering is rare in igneous rocks.

Plutonic rocks contain large crystals because they have cooled and solidified slowly, deep underground. (The name comes from Pluto, the Greek god of the underworld.) The slower the magma cools, the larger the crystals can grow and the coarser the texture. Plutonic rocks are also known as intrusive igneous rocks because they form by slow cooling within Earth. The formation of extrusive and intrusive rocks is illustrated in Figure 5-1.

Igneous rocks are classified on the basis of their crystal size and their mineral composition. **Felsic** rocks are light in color and relatively low in density. They are rich in quartz and feldspar, a light-color family of aluminum silicate minerals. **Mafic** rocks contain more iron and magnesium. Therefore, mafic rocks are usually more dense and darker in color than felsic rocks.

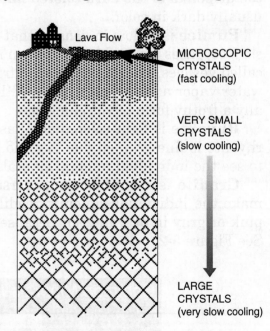

FIGURE 5-1. In general, slow cooling at depth produces larger crystals.

Objective

To identify igneous rocks based on their properties.

Materials

Set of rocks, Scheme for Igneous Rock Identification

Procedure

Take the rocks from your set and lay them on your desk in numerical order. As you read through this activity, compare your rock samples to the descriptions provided so that you will be able to identify your samples. To perform this lab you will need the Scheme for Igneous Rock Identification from the *Earth Science Reference Tables*. In the top part of the chart, igneous rocks are organized by two properties: grain (crystal) size and mineral composition. The coarse-grained rocks are at the bottom, just as they generally form within Earth. The felsic (light-colored) rocks are on the left. Toward the right they become more mafic (dark-colored). This chart contains a wide range of useful information. Locate on this chart each rock named in the descriptions below.

Rocks in the Scheme for Igneous Rock Identification

Obsidian is volcanic glass. The luster is glassy; it often breaks along curved surfaces (conchoidal fracture). Obsidian cools so rapidly that visible crystals cannot form. Due to the dispersal of the dark-colored minerals, both mafic and felsic varieties of obsidian are usually dark in color.

Pumice is so low in density that some samples float on water. As magma rises to the surface, gases trapped in the molten rock expand to form tiny pockets. These openings are called vesicles. The gases released from the magma as it comes to the surface are mostly water vapor and carbon dioxide. Although it does not look like obsidian, pumice is actually a frothy form of volcanic glass.

Rhyolite is light-colored and has a mineral composition similar to granite. However, rhyolite is fine-grained due to rapid cooling and solidification. Magnification is required to see the individual mineral crystals.

Granite is a light-colored, coarse-grained plutonic rock. The large grains (crystals) make the individual minerals readily visible without magnification. Granite is usually pink or gray in overall color because of the large quantity of felsic minerals it contains. See Figure 5-2.

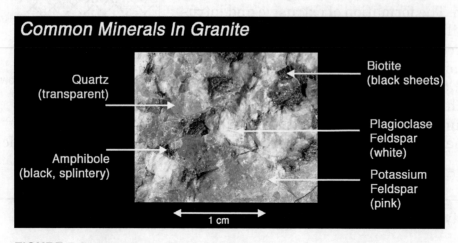

FIGURE 5-2. The primary minerals in granite.

Pegmatite is a very coarse-grained igneous rock. It is usually light in over all color, like granite. But the crystals are larger than 1 cm across. Pegmatite actually cools relatively quickly, not far below the surface. But the crystals grow very large because of the concentration of water in the magma.

Scoria is a rock full of larger vesicles (gas pockets), formed as gases expand within the cooling lava. In scoria the gas holes (vesicles) are big enough to be clearly visible. Scoria, like pumice, is a vesicular igneous rock.

Basalt is a fine-grained, dark-colored volcanic rock. Because it forms near the surface, basalt is composed of very fine crystals. It is rich in iron and magnesium minerals, so it is dark in color and relatively dense.

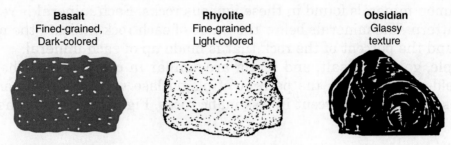

Basalt
Fined-grained,
Dark-colored

Rhyolite
Fine-grained,
Light-colored

Obsidian
Glassy
texture

FIGURE 5-3. Textures of igneous rocks.

Gabbro is a dark (mafic) plutonic rock. It is coarse-grained, like granite, but, because it is rich in iron and magnesium minerals, it is dark in color. Gabbro is the coarse-grained equivalent of basalt.

Use the Scheme for Igneous Rock Identification in the *Earth Science Reference Table*s to answer the following questions about igneous rocks.

1. What name is applied to a coarse-grained, felsic igneous rock? _____

2. What rock has the same composition as granite, but smaller crystals? _____

3. Name a coarse-grained, dark-colored igneous rock. _____

4. What feature of pumice and scoria makes them vesicular?

    _____

5. What two rock types are neither felsic nor mafic? (These two rocks have an intermediate composition and color.)

    _____

6. How do the two kinds of rock in question 5 differ from one another?

    _____

7. With respect to Earth's surface, where do extrusive rocks cool and crystallize? (That is, where do they solidify?)

    _____

8. What is the texture of the very finest grained igneous rocks called?

_____

9. What do gabbro and basalt have in common?

_____

10. How can you tell that gabbro crystallized deeper underground than basalt did?

_____

The lower portion of the Scheme for Igneous Rock Identification helps you to identify the most common minerals found in these igneous rocks. Each mineral is represented by a different pattern. The minerals below the name of each rock indicate the minerals usually present and the percent of the rock that is made up of each mineral.

For example, gabbro, basalt, and scoria are similar in composition. They all contain plagioclase feldspar. However, the percent of plagioclase varies from 0 percent in very mafic samples, to about 55 percent in less mafic rocks. Figure 5-4 shows the first step in

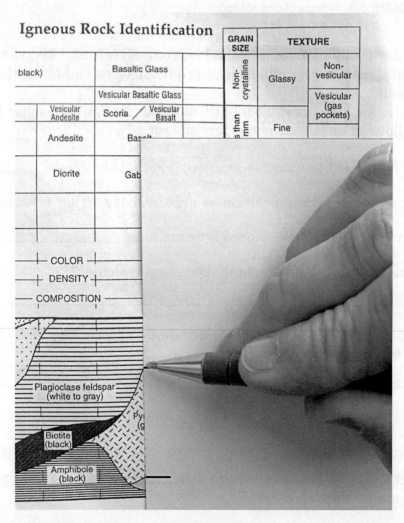

FIGURE 5-4. Marking the percentage of pyroxene in gabbro or basalt.

determining the typical pyroxene content of basalt and gabbro. Line up the edge of a piece of paper with the rock names and mark the top and bottom of the pyroxene section. Move the marked paper to the scale of percentages on the left side of the chart and read the number.

11. What minerals are usually found in granite?

_____

12. What is the percentage of plagioclase in a typical granite?

_____

13. List the four minerals usually found in basalt and the percentage of each mineral.

_____

_____

14. What two minerals tend to be found in the greatest range of igneous rocks?

_____ and _____

15. What igneous rock is generally composed of just one mineral?

_____

16. Unlike some other rocks, igneous rocks such as granite and gabbro are never composed of rounded particles. What are igneous rocks composed of?

_____

17. Do igneous rocks usually have a layered texture? _____

18. What coarse-grained rock is similar in composition to basalt? _____

19. What is the difference between granite and rhyolite?

_____

20. Unlike other kinds of rocks, igneous rocks form from what substance?

_____

Use your igneous rock samples to complete the table below. Some rocks may appear in the table more than once.

Sample Number	Color (Light or dark)	Grain Size (Coarse, fine, glassy)	Special Texture?	Name
1				
2				
3				
4				
5				
6				
7				
8				
9				
10				
11				
12				

When you have finished, please return a complete set of rocks. If the set is incomplete or you have extra items, let your teacher know.

CHAPTER 5—LAB 2: SEDIMENTARY ROCKS

Introduction

Unlike for igneous rocks, no single definition can be used to distinguish all sedimentary rocks from those that are not sedimentary. Most sedimentary rocks are composed of the weathered remains of other rocks that have been compressed and cemented in layers. Other sedimentary rocks are left behind when seawater evaporates or when organic remains are compacted and/or cemented by mineral material. Fossil remains of prehistoric life are found almost exclusively in sedimentary rocks.

Sedimentary rocks are classified into three groups on the basis of their origin; that is, how they formed. These three groups are clastic (fragmental), crystalline (mostly chemical precipitates), and bioclastic (organic).

CLASTIC (fragmental) rocks are the most common sedimentary rock. They are made from fragments of rocks that were weathered, eroded, and deposited as sediment. The sediment was compressed and cemented to form new rock. Unlike sediment, sedimentary rock is harder because it has been compressed and cemented. Compression is caused by the weight of material deposited on top of the rock layers. Silica (quartz), calcite (limestone), and clay are three common rock-forming cements. These cements are usually deposited by water seeping through the sediments. The clastic rocks are classified on the basis of the size of the grains of sediment. The following are clastic sedimentary rocks. (See Figure 5-5.)

SHALE is made of clay fragments less than 0.004 mm, so shale feels smooth and breaks into thin layers

SANDSTONE is composed of gritty sand grains from 0.004 mm 0.06 mm.

CONGLOMERATE contains pebbles greater than 2 mm in diameter cemented together.

FIGURE 5-5. Three clastic sedimentary rocks.

- **Shale** is composed of clay particles so small that they cannot be seen without magnification. Shale feels smooth and breaks into thin layers. Although dominated by the clay family of minerals, other minerals can be present if the particles are clay size.
- **Siltstone** is made of slightly larger particles, but it sometimes breaks into thin layers.
- **Sandstone** contains sand particles large enough to feel gritty. Although sandstone may or may not show layering, it usually breaks into irregular fragments.
- **Conglomerate** may look like a piece of cement containing pebbles. It is composed of pebbles, cobbles, or larger rock, rounded particles (fragments) of sediment held together by natural cement.
- **Breccia** is similar to conglomerate, except that its fragments are angular.

Most **CRYSTALLINE** sedimentary rocks are made of soluble materials that precipitated from seawater as the water evaporated. Therefore this group of rocks is sometimes known as the evaporites. Unlike the other sedimentary rocks, most rocks in this group are composed of relatively soft, intergrown crystals. However, you should remember that most rocks composed of intergrown crystals, such as igneous rocks, are not sedimentary rocks.

As ocean water evaporates, a variety of salts are left behind. **Rock salt** is the first and most abundant compound to precipitate. Sodium chloride (table salt), the mineral halite, is the essential mineral in the sedimentary rock called rock salt. Rock salt is followed in precipitation by other salts. **Rock gypsum** is one of the materials deposited later if evaporation continues. This kind of precipitation is now occurring in the Persian Gulf of Asia, and in the Great Salt Lake in Utah. Underground beds of rock salt in western New York State show that this part of North America was covered by an inland sea or gulf that was rapidly evaporating millions of years ago.

- **Dolostone** forms by a chemical reaction of limestone with seawater. Magnesium is added to the mineral calcite in limestone to change it to dolostone.

BIOCLASTIC (organic) sedimentary rocks are made from the remains of plants and animals. They are called organic because the rocks are made from material that was once alive, and because they all contain carbon.

- **Coal** is composed of the remains of plants that lived in tropical swamps millions of years ago. The plant material fell into water where it could not decay as quickly as it accumulated. Compression by burial turned these remains into peat, then lignite, and then into bituminous coal, which are relative low in density. Deeper burial may produce anthracite, commonly called hard coal, because it is harder and more dense than other forms of coal.
- **Coquina** is a variety of limestone composed entirely of bits of seashells cemented by a calcite matrix.
- Natural **chalk** is also composed of the remains of very tiny marine animals that are too small to be easily visible.
- **Limestone** is a sedimentary rock composed primarily of the mineral calcite. Most limestone was formed from shells and other parts of marine organisms. If shells have been abraded into sand-size particles composed of calcite, fragmental limestone is the result. Some calcites may have formed by chemical precipitation. Thus, limestone is a sedimentary rock that has sometimes been classified into any of the three above mentioned sedimentary groups.

Objective

To identify sedimentary rocks based on their properties.

Materials

Set of sedimentary rocks, Scheme for Sedimentary Rock Identification chart

Procedure

Obtain a set of sedimentary rocks. Place them in numerical order on your desk as you read this paper. If your set is incomplete, or if you have extras, please tell your teacher.

Use the descriptions above and the Scheme for Sedimentary Rock Identification to identify your sedimentary rock samples. (Some rocks may be used more than once.)

	Rock Name	Distinguishing Characteristics
1.	_____	_____
2.	_____	_____
3.	_____	_____
4.	_____	_____
5.	_____	_____
6.	_____	_____
7.	_____	_____
8.	_____	_____
9.	_____	_____
10.	_____	_____
11.	_____	_____
12.	_____	_____

Next, circle the names of all of the rocks in your set that are monomineralic. (That is, they are generally composed of just one mineral or one kind of substance.).

Please be sure to return a complete set of rocks, ready for the next group. If your set has broken samples, is incomplete, or has extra samples, please tell your teacher.

Wrap-Up ▐▐▐▶

Use the Scheme for Sedimentary Rock Identification to answer the following questions. The right side of the Scheme for Sedimentary Rock Identification shows symbols often used to represent various sedimentary rocks on diagrams and maps. In most cases, the symbol shows something about the texture of the rock. For example, shale usually breaks into thin horizontal layers, while sandstone is made of little grains of sand.

1. What is coal made from? _____

2. What clastic rock is composed of the smallest grains of sediment?

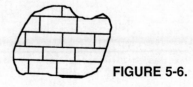

FIGURE 5-6.

3. What rock is often represented in diagrams by this pattern? (Figure 5-6)

_____

4. What rock is made of compressed and hardened clay? _____

5. In what way is breccia unlike conglomerate?

6. What is the largest diameter of rock fragment generally found in siltstone?

7. What common sedimentary rock is composed mostly of the mineral calcite?

8. What three minerals or mineral families dominate all the clastic sedimentary rocks?

9. If a rock is made of grains mostly about 1 mm across, what kind of rock is it?

10. Although Earth is mostly igneous rock, most of the bedrock we see at the surface is sedimentary. Why?

We can usually identify a rock as sedimentary because it has one or more of the following characteristics:

1. It is composed of rounded fragments compressed and cemented together.

2. It is layered, although the layers may be too thick to show in a small sample.

3. It contains fossils. (Fossils are not found in all sedimentary rocks; however, nonsedimentary rocks very seldom contain fossils.)

CHAPTER 5—LAB 3: METAMORPHIC ROCKS

Introduction

The word metamorphic means changed in form. Intense heat and/or pressure without melting change an older rock to a new metamorphic rock. Most metamorphic rocks begin as igneous or sedimentary rocks. Metamorphism causes new minerals to form, crystals to grow, structures such as sedimentary layers to become distorted, and density to increase.

Figure 5-7 shows that as sediments such as clay are buried deeper within Earth, heat and pressure increase, causing the sediments to change into shale, a sedimentary rock. The shale is changed by deeper burial through slate, phyllite (FILL-ite), schist, and gneiss (NICE). Still higher temperatures and deeper burial could cause the rock to melt. Then, cooling and crystallization would produce an igneous rock.

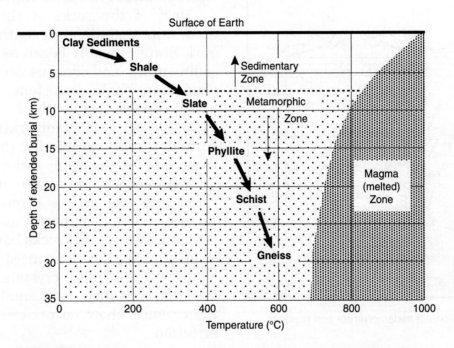

FIGURE 5-7. Changes in rock caused by heat and pressure at increasing depth.

Metamorphic rocks follow a continuum from relatively slightly changed to higher grade rocks that have undergone more extensive alteration. The greater the changes caused by metamorphism, such as foliation and the formation of new minerals, the higher the grade of metamorphism.

Many metamorphic rocks display both mineral crystals and layering. This layering may be the result of crystal growth, and not the remains of original sedimentary bedding. The alignment of mineral crystals is known as foliation. Foliation is a common property of metamorphic rocks, as shown in Figure 5-8.

FIGURE 5-8. Foliation is layering caused by alignment of mineral crystals, especially mica.

FIGURE 5-9. Banding is a result of minerals separating into layers during metamorphism.

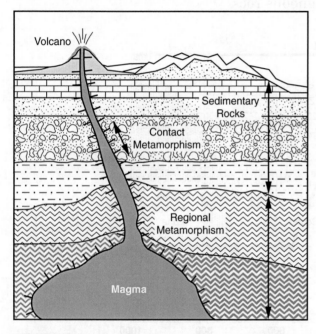

FIGURE 5-10. Contact metamorphism and regional metamorphism.

Banding is a coarser kind of layering, resulting in the separation of minerals into bands, often forming light and dark layers as shown in Figure 5-9

Figure 5-10 illustrates the two kinds of metamorphism. **Contact metamorphism** occurs in a narrow zone next to an igneous intrusion. Rocks in contact with the intrusion are baked by the heat they absorb, causing chemical changes that can create new minerals. Usually the metamorphic grade (degree of change) decreases with distance from the heat source. Most contact metamorphic rocks, such as hornfels, do not show foliation or banding. However, when a large mass of rock is buried many kilometers within Earth, heat and pressure can cause widespread **regional metamorphism**.

Most of the rocks of the Adirondack Mountain region and in southeastern New York State probably began as a mixture of sedimentary and igneous rocks, including sandstone, shale, limestone, granite, and basalt. Burial at a depth of as much as 20 to 30 km, where the temperature and pressure are high, gradually changed these rocks into metamorphic rocks.

Metamorphic rocks are usually separated into two groups according to their texture (foliated or nonfoliated) and composition. The foliated rocks have a streaked or layered appearance caused by the parallel growth of mineral crystals. The nonfoliated rocks contain very small crystals, or the crystals show no preferred alignment direction.

Foliated Metamorphic Rocks

The following five metamorphic rocks may be similar in composition, but they differ in the amount of metamorphic alteration.

- **Slate** looks like shale, but it is harder and more dense. Unlike shale, slate often breaks along the mineral foliation direction rather than along the original sedimentary bedding. It is usually gray, black, or brown.
- **Phyllite** (FILL-ite) looks similar to slate; however, phyllite is shiny due to the growth of tiny mica crystals. The layering in phyllite is sometimes wavy.
- **Schist** contains mica crystals large enough to be easily visible. Nonfoliated minerals, such as quartz and feldspar may also be present in a few veins or bands, although the mica foliated texture still dominates.

- **Gneiss** (NICE) has a banded look with alternating layers of light-colored minerals, such as quartz and feldspar, and dark minerals such as amphibole and pyroxene. Schist changes to gneiss when most of its flat mica crystals change to other minerals including feldspar. Some gneisses do not show banding and look almost like igneous granite. However, you know it is metamorphic because crystals show a foliation direction.

Increased heat and pressure will cause gneiss to melt and then solidify, changing the metamorphic rock into magma and then an igneous rock.

Nonfoliated Metamorphic Rocks

The nonfoliated metamorphic rocks have variable origin and composition.

- **Marble** is metamorphosed limestone (mostly calcite). It is usually white, gray, or other light color. Some examples are foliated. Powdered calcite mineral, made by hitting marble with a rock hammer, will bubble with a strong acid.
- **Quartzite** is metamorphosed sandstone. It is usually light in color and it commonly has a glassy luster on freshly broken surfaces. Because it is so hard, quartzite weathers slowly to make smooth, rounded surfaces and pebbles.
- **Metaconglomerate**, unlike the sedimentary conglomerate from which it forms, often splits through the pebbles. The pebbles may be squashed or elongated by the combination of heat and pressure deep within Earth.

Objective

To identify metamorphic rocks based on their properties.

Materials

Set of metamorphic rocks, Metamorphic Rock Identification chart

Procedure

Place your set of rocks in numerical order on the desk in front of you. Identify each of your metamorphic rocks using the *Earth Science Reference Tables* and the descriptions above.

Complete the grid below based on your identification of the metamorphic rock samples. Rock names may be used more than once.

Sample Number	Texture	Composition	Rock Name
1			
2			
3			
4			
5			
6			
7			
8			
9			
10			
11			
12			

Return your samples. Please let your teacher know if your set is incomplete or if you have extra items.

Wrap-Up

1. What are metamorphic rocks made from?

_____

2. What two conditions change a rock into a metamorphic rock?

_____

3. What mineral feature forms and grows larger with increasing metamorphism?

_____

4. What are foliation and banding?

_____

_____

5. What are the two types of metamorphism?

_____

_____

6. Which kind of metamorphism occurs next to an intrusion of molten magma?

_____

7. If clay is buried very deep within Earth, it can progress through a series of rocks of increasing metamorphic grade. Name the rocks in this series.

_____

8. What process is involved in the formation of all igneous rocks, but not in the formation of metamorphic rocks?

_____

_____

9. What are some of the changes that occur when a sedimentary rock transforms into a metamorphic rock?

_____

_____

10. In what parts of New York State is metamorphic bedrock most common?

_____

Base your answers to questions 11–20 on the Scheme for Metamorphic Rock Identification in the *Earth Science Reference Tables*.

11. Which foliated metamorphic rock has the smallest crystals? _____

12. What mineral family is often found in most foliated metamorphic rocks?

_____

13. What sedimentary rocks can change into marble? _____

14. Of the metamorphic minerals shown in the chart under Composition, which one seems to require the most heat and pressure, so it is generally seen only in gneiss or high-grade schist?

_____

15. What mineral family can the mica minerals change to when gneiss is formed?

_____

16. In what two ways does metaconglomerate often differ from a sedimentary conglomerate?

_____

17. Which type of metamorphic rock is usually foliated and banded? _____

18. Only one metamorphic rock is always formed by contact metamorphism. What is it?

_____

19. What group of minerals usually gives schist a strong foliation? _____

20. Which kind of metamorphic rock is found in the central Adirondacks?

(This question probably requires you to use a different page in the *Reference Tables*.)

_____

🌐 CHAPTER 5—SKILL SHEET 1: THE ROCK CYCLE

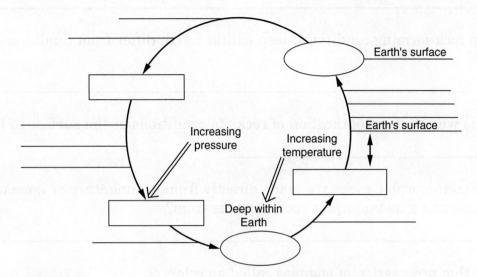

FIGURE 5-11.

Directions

 A. In the appropriate boxes in Figure 5-11, write the five forms of rock materials. You should use the *Earth Science Reference Tables* Rock Cycle in Earth's Crust diagram.

 B. On the lines in Figure 5-11, write the processes that change one form of rock material to another. Be sure that these changes occur in the proper sequence.

 C. Draw arrows on Figure 5-11 to show alternative paths that rock materials may follow, in addition to the path around the outside of the circle. Label these new paths with the changes they represent.

Answer the following questions.

 1. What determines whether we classify a rock as igneous, sedimentary, or metamorphic?

  _____

 2. What two Earth materials around the outside of the circle are not actually rocks?

  _____

 3. What name is applied to magma when it is extruded at Earth's surface? _____

 4. Explain why igneous or metamorphic rocks cannot change directly into sedimentary rocks.

  _____

5. What step is necessary in the formation of all igneous rocks, but not to sedimentary or metamorphic rocks?

_____

6. How do rock-forming conditions deep within Earth differ from conditions at Earth's surface?

_____

7. What do we call the modification of rocks to conditions at the surface of Earth?

_____

8. Not all metamorphic rocks are made directly from sedimentary or igneous rock. What else can a metamorphic rock be made from?

_____

9. Why is this progression of changes called a cycle?

_____

10. According to Figure 5-11, which material is formed at Earth's surface?

_____

11. According to both rock cycle diagrams, what are almost all rocks made from?

_____

12. Name two different rocks formed by the process indicated by the arrow at the top labeled Oxygen, Water, and Organic Materials.

_____

Chapter 6
Managing Natural Resources

CHAPTER 6—SKILL SHEET 1: NATURAL RESOURCES

People use a wide variety and great quantities of natural resources. Your teacher may want you to complete the following in cooperative groups. If so, one completed copy per group is enough; however, include the names of all in the group. Ask your teacher for help if you need it.

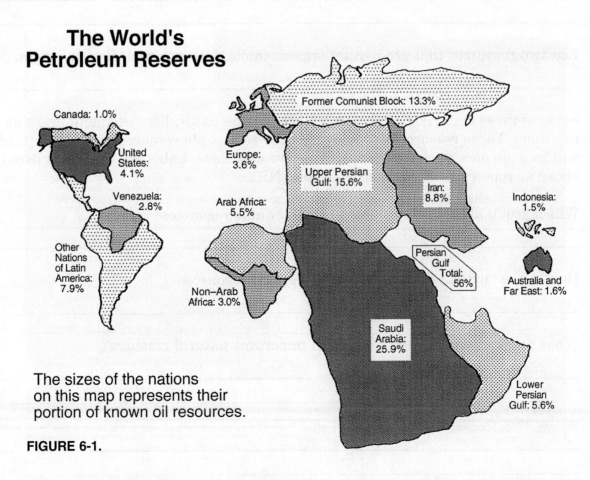

The World's Petroleum Reserves

Canada: 1.0%

United States: 4.1%

Venezuela: 2.8%

Other Nations of Latin America: 7.9%

Former Comunist Block: 13.3%

Europe: 3.6%

Upper Persian Gulf: 15.6%

Iran: 8.8%

Indonesia: 1.5%

Arab Africa: 5.5%

Persian Gulf Total: 56%

Australia and Far East: 1.6%

Non–Arab Africa: 3.0%

Saudi Arabia: 25.9%

Lower Persian Gulf: 5.6%

The sizes of the nations on this map represents their portion of known oil resources.

FIGURE 6-1.

1. The map in Figure 6-1 shows the portion of Earth's known crude oil reserves in various countries. This kind of map is a visual representation of data. What does

this map tell you about the distribution of crude oil reserves and the consequences it has for the United States?

_____

_____

_____

_____

_____

2. List five common materials you use in your everyday life that are mostly of organic (biological) origin.

_____

_____

3. List five common materials that are mostly of geological origin.

_____

_____

4. List two resources that are neither organic (biological) nor geological in origin.

_____

_____

5. Some of these materials are being replaced in the natural environment, even as we use them. These resources are known as the renewable resources. Other materials will be gone once we have used up what we now have. Label the resources listed above as renewable (R) or nonrenewable (NR).

6. Which group above includes more nonrenewable resources?

_____

7. How can we extend the availability of natural resources?

_____

8. What happens when we run out of an important natural resource?

_____

_____

_____

_____

Chapter 7
Earthquakes and
Earth's Interior

CHAPTER 7—SKILL SHEET 1: SEISMIC SKILLS

Earthquakes occur when stresses within Earth's crust become so great that the ground breaks. The breakage occurs along zones of weakness. The vibrations we feel as earthquakes are caused by energy waves that radiate from these breaks. The huge Indonesian earthquake of December 26, 2004 occurred when rocks along a fault more than 1000 km (600 miles) long in the Indian Ocean suddenly shifted about 20 meters (approximately 70 feet).

FIGURE 7-1. A fault along a highway in Arizona; the displacement along this fault is about 0.5 meter.

In 1902, the Italian geophysicist Giuseppe Mercalli published a relative scale for measuring earthquakes. This scale was based on firsthand observations or published reports of the effects of earthquakes. On the Mercalli scale, an earthquake of intensity I would be so slight that very few people would feel it. The most intense seismic events are rated XII, and may cause total destruction of structures that are not reinforced by steel frames. This is a relative scale because it is not a direct measurement of the shaking or the energy of the event.

In 1935, Dr. Charles Richter at the California Institute of Technology proposed a scale of absolute magnitude. On this scale, an earthquake of magnitude 1 could be observed only with a sensitive instrument, such as a seismograph. On this scale, for each increase of 1 unit, the ground shakes 10 times as much. For this reason it is known as a logarithmic scale. In fact, the change in total energy released is even greater: about 30 times as much with each one-step increase in magnitude.

Seismic magnitude is determined by measuring ground motion with an instrument called a seismograph (seismometer). Mechanical seismographs use a heavy weight suspended on a spring. Figure 7-2 shows the concept. When the ground moves suddenly, the weight tends to remain still. A pen attached to the weight also stays still while the ground and the recording paper move. (The paper is mounted on a slowly rotating cylinder.) During an earthquake, the pen traces a zigzag path on the moving paper. This record of the movement is known as a seismogram. The stronger the earthquake, the farther the tracing moves up and down. In modern seismographs, the weight is suspended by electromagnetism, which can be used to amplify weak signals.

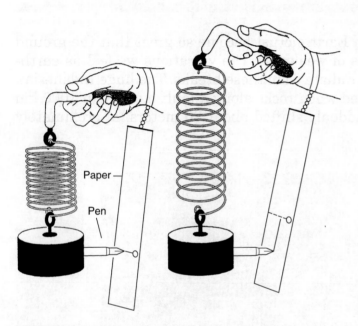

Instruments used for the original Richter scale were calibrated to the geology and buildings found in Southern California. Their measurements were not useful for seismic events that occurred in other parts of the world. Therefore, the moment-magnitude scale has largely replaced the Richter scale.

FIGURE 7-2. When the hand is held still, the paper is still, and the suspended weight is still. When the hand and the paper attached to move up suddenly, the spring stretches, the weight does not move, and the pen traces a line on the moving paper.

To avoid confusion, the moment-magnitude scale has been adjusted to conform to Richter magnitudes as much as possible. Some people still call it the Richter scale.

The greatest earthquake in modern history occurred in the Pacific Ocean off Chile in 1960. It had a magnitude of about 9.5. That seems to be the limit of stress terrestrial rocks can undergo before they break. The largest recent seismic event was the Indonesian earthquake of 2004: magnitude 9.1–9.3.

1. What causes earthquakes?

_____

2. What geologic structures are associated with earthquakes? _____

LABORATORY MANUAL

3. What instrument is used to record and measure earthquakes

_____

4. How much more ground movement does a magnitude 6 earthquake produce than a magnitude 5?

_____

_____

5. How much greater is the ground shaking in a magnitude 7 seismic event than a magnitude 4 event?

_____

6. Why is it easier to assign Mercalli intensity to historic earthquakes than it is to give them a Richter-type magnitude?

_____

The place underground where the rock starts to break is known as the hypocenter (focus) of the earthquake. Directly above the focus, on the surface, is the epicenter. People at or near the epicenter will feel an earthquake most strongly.

Energy radiates from the focus as vibrational waves. The speed of these waves is directly proportional to the rigidity of the rock through which they travel. The study of vibrations from underground explosions and earthquakes has given geologists important information about Earth's interior.

Earthquakes generate surface waves and two kinds of energy waves that travel through Earth (body waves): primary and secondary waves. These waves travel through the solid Earth much like sound and light waves radiate through air. They are useful in locating earthquake epicenters. Each type of wave has its own special characteristics.

Primary (P) waves travel fastest. P-waves are longitudinal waves, like sound waves, because the waves pass as a series of compressions and expansions. P-waves cause Earth to vibrate back and forth in the direction the wave travels, in a push-pull motion. Like sound, P-waves can travel though solids and liquids.

Secondary (S) waves are a little slower than P-waves. As you can see in Figure 7-3, the S-wave will arrive after the P-wave. Like light, S-waves are transverse waves that cause side-to-side vibrations, making the ground move perpendicular to the direction of travel. For this reason, S-waves are also known as shear waves. However,

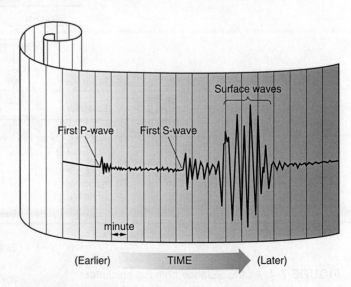

FIGURE 7-3. As the paper is drawn past the recording pen, a line is made to show ground movement. Notice the order of P-, S-, and surface waves.

S-waves cannot travel through liquids. The discovery of Earth's liquid outer core was made when it became clear that S-waves do not travel through this part of the center of Earth.

Surface waves are the last to arrive. (This is partly because they travel through less rigid, more flexible rocks near Earth's surface.) Like waves on water, surface waves involve a combination of longitudinal and transverse vibrations. Surface waves cause the most damage and loss of life.

7. What is directly below the epicenter of an earthquake?

_____

8. The rock of Earth becomes more rigid with depth. As P- and S-waves travel deeper within Earth, what happens to the speed of these energy waves?

_____

9. Which of the seismic waves causes the most damage to human-made structures?

_____

10. When an S-wave passes, in what way does the ground vibrate?

_____

11. Which seismic waves will not pass through a liquid? _____

Because P-waves travel faster than S-waves, the greater the distance between the epicenter and the recording station, the greater the time delay of the S-waves. For a nearby earthquake, the P- and S-waves arrive with little separation. However, for distant events, the time delay, or lag, will be longer. Consider Figure 7-4. P- and S-waves start out to-

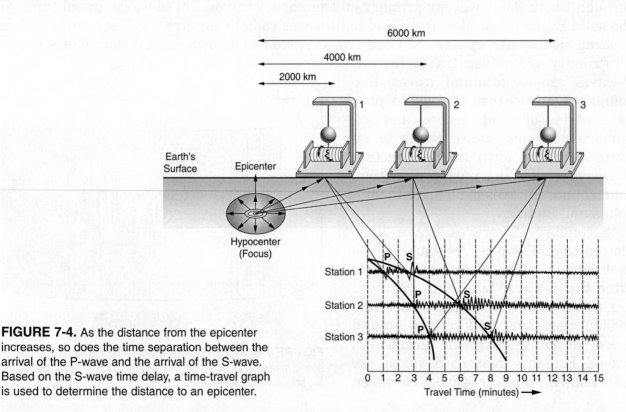

FIGURE 7-4. As the distance from the epicenter increases, so does the time separation between the arrival of the P-wave and the arrival of the S-wave. Based on the S-wave time delay, a time-travel graph is used to determine the distance to an epicenter.

gether at the epicenter. By the time they have traveled 2000 km, the S-wave is 3 minutes 20 seconds behind the P-wave. This specific time separation, or delay, occurs only at a distance of 2000 km. Therefore, the delay in the S-wave arrival can be used to determine the distance from the recording location to the earthquake epicenter.

If the distance between the epicenter and three widely separated recording stations is known, the location of the earthquake can be determined. If there are more than three recording stations, greater accuracy can be achieved.

Figure 7-5 shows three seismic recordings (seismograms) from the same earthquake detected at three different seismic stations. The smallest wiggles are called ambient "noise" because it is always present and does not indicate an earthquake. Passing traffic, waves on a shore, or people walking near the instrument can cause noise. Use this diagram to answer questions 12, 13, and 14.

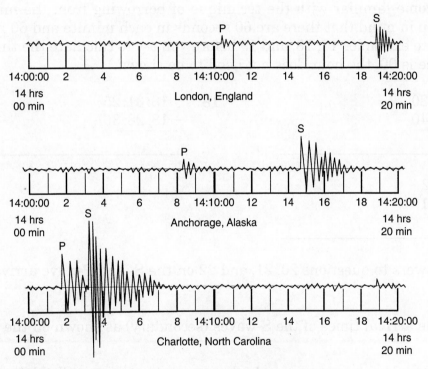

FIGURE 7-5. Three seismograms from a single earthquake.

12. At any one station, which causes the ground to vibrate more, P- or S-waves?

_____

13. How much time is represented by the distance separating one vertical line from the next?

_____

14. Why are the vibrations from this earthquake larger at Charlotte than they are in London?

_____

Records of P- and S-wave arrival times are written as HOURS:MINUTES:SECONDS in 24-hour military time. Therefore, 08:27:30 would mean 30 seconds past 8:27 in the morning, and 13:00:05 would be 5 seconds past 1 in the afternoon (12 noon + 1 hour = 13 hours).

15. In military time, when does the first class period begin in your school?

_____

16. The Indonesian earthquake of 2004 began on December 26 at 3 seconds past 4:52 P.M. New York time. How should this time of day be written in military time?

_____

In a few minutes you will calculate the time delay between P- and S-waves. However, you must first become familiar with the technique of borrowing from the minutes or hours columns. (Keep in mind that there are 60 seconds in each minute and 60 minutes in each hour.) Complete examples 17, 18, and 19 before you try to calculate the time delay. (If the hour difference is 00, the hour does not need to be shown.)

17. $10:18:30$
     $-\;10:11:40$

    _____

18. $18:31:25$
     $-\;18:28:36$

    _____

19. $10:01:22$
     $-\;9:56:31$

    _____

Base your answers to questions 20, 21, and 22 on the S- and P-wave arrival times shown in Figure 7-5.

20. Record the arrival times of the S-waves (secondary) as shown on the seismograms above.

S-waves: London _____ Anchorage _____ Charlotte _____

21. Next, record the arrival times of the P-waves at these stations.

P-waves: London _____ Anchorage _____ Charlotte _____

22. Calculate the S-wave delay in London, Anchorage, and Charlotte.

_____

Once the time difference between the P- and S-wave arrivals are known, the distance from the epicenter to any seismic station can be found. Use the Earthquake P-wave and S-wave Travel Time graph in the *Earth Science Reference Tables* to answer the following questions.

23. What quantity is measured along the horizontal axis of the P- and S-wave Travel-Time graph?

_____

24. The vertical axis measures what quantity?

_____

25. What is the interval between the smallest lines on the vertical axis?

_____

26. What is the interval between the smallest lines on the horizontal axis?

_____

27. Why is the S-wave line higher than the P-wave line?

_____

28. How long does it take a P-wave to travel 2000 km? (To the nearest 10 seconds.)

_____

29. How long does it take an S-wave to travel 2000 km?

_____

30. As the distance from the epicenter increases, what happens to the time lag between P-and S-waves?

_____

31. Just by looking at Figure 7-5, there are at least three ways you can tell that London is farther from the epicenter than the other two seismic stations. Give at least two.

_____

_____

_____

This delay allows you to determine the distance from the epicenter to each recording station, using the graph in the *Earth Science Reference Tables*. To understand this you should now lay a straight edge of a piece of paper along the vertical axis just as shown Figure 7-6

Carefully make a mark along the blank edge at 0 minutes and another at the 5-minute mark. (Mark your paper now as shown in Figure 7-6.)

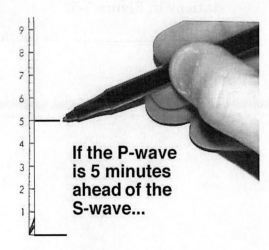

If the P-wave is 5 minutes ahead of the S-wave...

FIGURE 7-6. Marking the S-wave delay on a paper edge.

Next, move the paper to the place where there is a time difference of 5 minutes between the P- and S-wave lines. (See Figure 7-7.) Take care to keep the blank edge vertical.

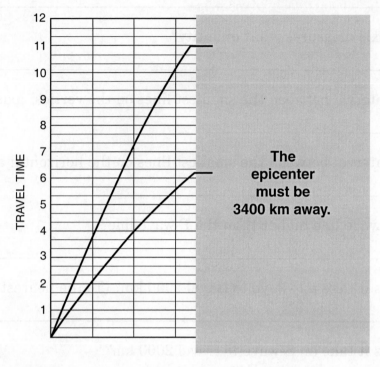

FIGURE 7-7. Using time delay to find epicenter distance.

32. At what distance will P-waves arrive 2 minutes ahead of S-waves?

_____

33. If the P-wave arrives 8 minutes before the S-wave, what is the distance to the epicenter?

_____

34. For this last question, you will need to refer to the numbers that you calculated in question 22. How far was this earthquake epicenter from the three recording stations in Figure 7-5?

London _____   Anchorage _____   Charlotte _____

CHAPTER 7—LAB 1: LOCATING EPICENTERS

Introduction

The epicenter of an earthquake is usually determined by examining seismograms from at least three recording stations. From these records, the distance to the epicenter of the earthquake from each of the recording stations can be determined. Circles drawn on a map around each of the seismic stations are used to locate the epicenter. In addition, the seismic recordings can be used to determine the time at which the earthquake took place and how powerful the earthquake was at its source.

Objective

To locate the epicenter of an earthquake.

Materials

Lab Sheets

Procedure

1. What is the time separation between the vertical lines in Figure 7-8?

   _____

   (Please note that the times on this chart are shown as Hours: Minutes: Seconds.)

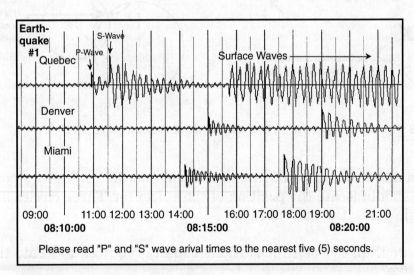

FIGURE 7-8. The first earthquake.

2. Which type of earthquake wave arrives first? _____

3. What wave type, because they move more slowly through Earth, always arrives after the P-waves? (It is usually more intense.)

Use the seismograms above and the Earthquake P-wave and S-wave Travel Time graph in the *Earth Science Reference Tables* to complete the table below. Please record all arrival times to the nearest 5 seconds

EARTHQUAKE #1 DATA TABLE

Seismic Station	P-wave Arrival Time	S-wave Arrival Time	S-wave Time − P-wave Time	Distance to Epicenter	P-wave Travel Time
Quebec					

The data that you entered in the table above can be used to locate the epicenter and to find the time at which this earthquake occurred at the epicenter. Figure 7-9 shows that with a single station, you can tell how far away the epicenter is, but you are not able to determine the direction to the epicenter.

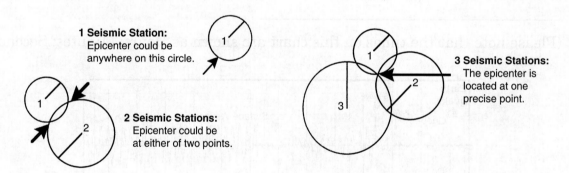

1 Seismic Station: Epicenter could be anywhere on this circle.

2 Seismic Stations: Epicenter could be at either of two points.

3 Seismic Stations: The epicenter is located at one precise point.

FIGURE 7-9. At least three seismic stations are needed to locate a single point.

A second station allows you to make another circle and determine two possible locations. The third circle should locate the position of the epicenter. Three circles are needed to intersect at a single point: the epicenter.

4. Why are more than two seismic stations usually required to locate the epicenter of an earthquake?

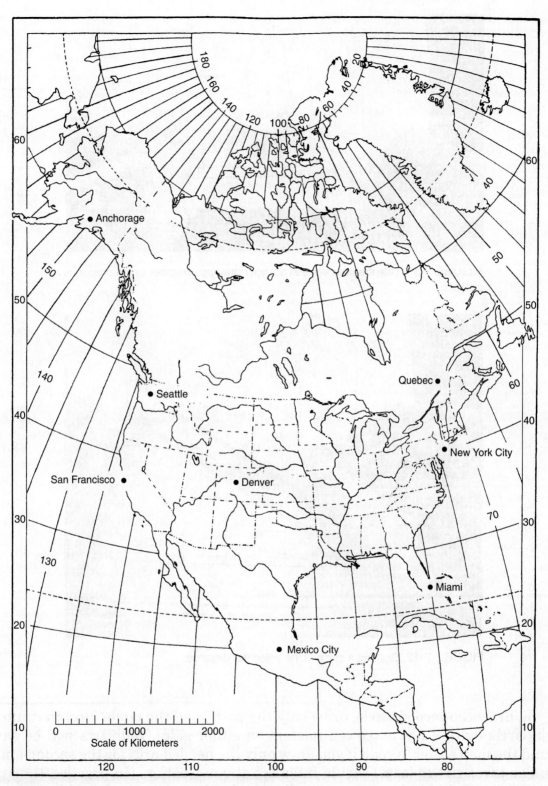

FIGURE 7-10.

To locate the epicenter of the earthquake, as shown on page 82, you will need a drawing compass, a pencil, and a copy of the earthquake travel time graph. You will find a map of North America in Figure 7-10. Use the map scale to stretch out the drawing compass to the proper distance from each recording station. (Follow the example in Figure 7-11.) Draw a circle (in pencil) at the proper distance around the first station, as shown in Figure 7-12.

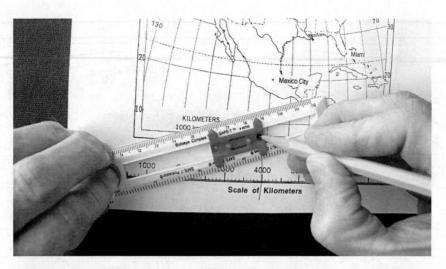

FIGURE 7-11. Use the map scale to set your drawing compass to the proper distance.

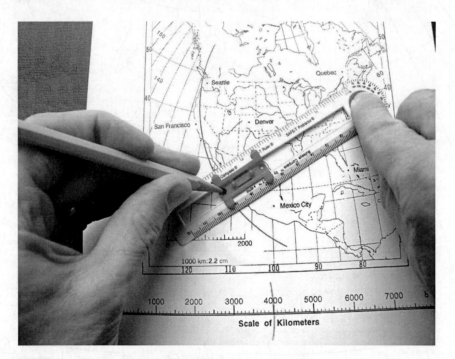

FIGURE 7-12. Drawing a circle at the epicenter distance.

When you draw the second circle, draw only the portion of the circle (arc) that intersects the first circle (with a few extra centimeters on either side). Your arcs will be easier to draw, and the map easier to read if you draw only the needed arcs for the second and third circles. The three circles (arcs) should intersect at one location. (In practice, the arcs seldom come to single point. If you get a small triangle, the epicenter should be at about the center of the triangle.)

The actual time at which the earthquake took place at the epicenter is called the origin time. Once the distance from the epicenter to any one recording station is known, it is possible to use a single seismogram find the origin time of the earthquake. You can check the origin time by using P-waves from other stations or by using any S-waves.

5. Based on the travel time graph, how long does it take a P-wave to travel 4000 km?

   _____

6. If the P-waves from this earthquake 4000 km away arrived at a recording station at exactly 12:00:00 (noon), when did the P-wave start its journey? (That is, what was the earthquake origin time?)

   _____

7. *a.* Time the P-wave arrived at Quebec: _____

   b. Epicenter distance from Quebec: _____

   c. Travel time for Quebec P-wave: _____

   d. Origin time (arrival time − travel time): _____

The table below provides data for a second earthquake. Complete this table and use the data to draw circles to locate the epicenter. Use the same map that you used for the first earthquake. (Once again, draw only the parts of the second and third circles that you really need.) Clearly label on the map, the location of epicenter the first (#1) and the second (#2) earthquake.

EARTHQUAKE #2 DATA TABLE

Seismic Station	P-wave Arrival Time	S-wave Arrival Time	S-wave Time − P-wave Time	Distance to Epicenter	P-wave Travel Time	Origin Time
Seattle	13:08:10	13:10:50				
Denver	13:07:35	13:09:50				
Anchorage	13:11:50	13:17:15				

8. Which column in the table above should be about the same value for all three entries?

   _____

9. Briefly explain how you can determine the distance to an earthquake epicenter from the records of a single seismic recording location? (Please number the steps of your procedure, #1, 2, 3, etc.)

   _____

   _____

   _____

   _____

Figure 7-13 contains seismograms from five recording stations. Use this information to locate the epicenter on the map and to determine the origin time of this event. Label the epicenter #3. Please draw only the useful parts of all five circles. Even though three should be enough to show the location of the epicenter, circles 4 and 5 are to check for accuracy.

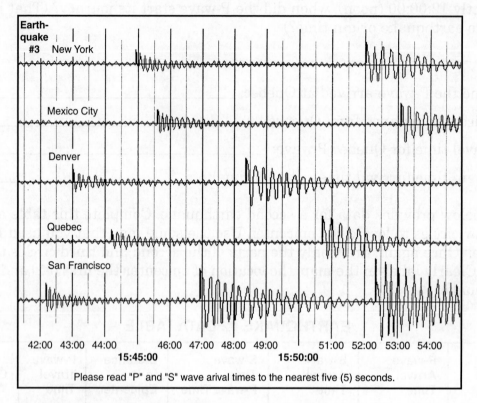

FIGURE 7-13. The third earthquake.

EARTHQUAKE #3 DATA TABLE

Seismic Station	P-wave Arrival Time	S-wave Arrival Time	S-wave Time – P-wave Time	Distance to Epicenter	P-wave Travel Time	Origin Time

10. What is your best estimate of the origin time of the third earthquake?

_____

(Be sure the epicenter locations are marked on the map: #1, #2, and #3. Then continue with the lab.)

Wrap-Up

11. How many recording stations are required to locate an epicenter?

_____

12. In what three ways do P-waves differ from S-waves?

_____

_____

_____

13. To locate an epicenter, subtract the

a. _____ arrival time from the

b. _____ arrival time. Then use the Travel Time graph printed in the

c. _____ to find the corresponding

d. _____ to the epicenter. Do the same for at least

e. _____ more seismic stations. Around each seismic recording station, draw a

f. _____ at the proper distance. Where the three circles

g. _____ is the epicenter. To find the origin time, subtract the

h. _____ for any P- or S-wave from the

i. _____ of that wave. That's when the earthquake occurred at the epicenter.

14. Define latitude. _____

15. Define longitude. _____

16. From the map, list the terrestrial coordinates of the three epicenters you located in this lab. (Be sure to label each with the angle and the direction: such as 45° S, 127° E.)

Event #1   **Event #2**   **Event #3**

Latitude: _____   Latitude: _____   Latitude: _____

Longitude: _____   Longitude: _____   Longitude: _____

17. You cannot locate the epicenter of an earthquake with seismic records from just one recording station. What information can you determine about the earthquake at its source from a single seismic station? (List only properties of the event at the epicenter. There are at least three answers to this question. You must list two for full credit.)

a. _____

b. _____

(Additional answers are optional.)

c. _____

CHAPTER 7—SKILL SHEET 2: A JOURNEY TO THE CENTER OF THE EARTH

No person has ever been to Earth's center, but scientists do have a good idea of what they probably would find. Like detectives, scientists must conduct investigations and interpret the evidence. The clues about the inner Earth are many.

Of course, the most direct way to find out about Earth's interior would be to retrieve rocks from deep underground. That has actually been tried. In the Russian Arctic, scientists have drilled the world's deepest borehole, which reaches to a depth of about 16 km.

Deep drilling in many places has shown that metamorphic and igneous rocks usually underlie the sedimentary rocks that cover most of Earth. The continents contain a core of granitic rocks. This includes granite and rocks of similar mineral composition. Below the continental granitic rocks, as well as below oceanic sediments, is a darker, more dense igneous rock, similar in composition to basalt.

Seismic measurements have shown that earthquake waves increase in speed just below the boundary called the Mohorovicic discontinuity. This interface is also called the Moho. The speed of seismic waves changes with changes in rock type. For this reason, many geologists believe that the Moho is the region where basalt changes to a more dense, high-pressure, basaltic rock known as eclogite. The depth to the Moho is about 5 to 10 km under the oceans and averages 40 km under the continents. Under mountains the crust can be as much as 80 km thick. The layer above the Moho is known as the crust, and the layer below this interface is Earth's mantle. Figure 7-14 shows the inferred layers near Earth's surface in a highly simplified diagram.

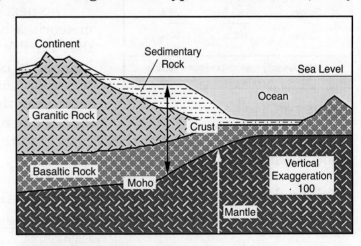

FIGURE 7-14. A profile of the geosphere.

1. What is the name of the boundary at the bottom of the crust?

_____

2. What kind of rock makes up the continents? _____

3. What layer of Earth is immediately below the crust? _____

4. Where is Earth's crust thickest?

_____

5. How is granite different from basalt? (Give at least two major differences.)

_____

_____

_____

Scientists have formulated their ideas about the center of Earth from the study of meteorites. To interpret this evidence, geologists make inferences. Most scientists infer that Earth and the solar system originated from a great cloud of dust and larger particles. The material was drawn together by its own gravity.

If Earth was formed from meteorites, then Earth's average composition should be similar to the average composition of meteorites. In fact, the crust and the mantle (at least what we think is in the mantle) are similar in composition to stony meteorites. However, the crust and mantle contain relatively little iron. Figure 7-15 shows that many meteorites are rich in iron. If meteorites represent the composition of Earth, where is Earth's missing iron?

The next step is deduction. If iron is common in meteorites, but not in the crust or mantle, it is likely that Earth's core is rich in iron. Being very dense, iron would be drawn by gravity toward Earth's center. Additionally, calculations of the mass of Earth show that the average density of Earth is greater than the density of rocks found in the crust as well as rocks of the mantle. All of these clues lead to the interpretation that Earth's core is made of a high density material, mostly iron.

TERRESTIAL METEORITES*		
Class	Falls (%)	Finds (%)
Iron	5	67
Stony-iron	1.5	7.5
Stony	93.5	25.5
Totals	100%	100%
* Includes only meteorites found outside Antarctica		

FIGURE 7-15.

6. Why do scientists study objects from outer space to investigate Earth's interior?

_____

7. What percentage of the meteorites that fall to Earth are stony meteorites? _____

8. What percentage of the meteorites found are the stony meteorites? _____

9. Suggest a reason why iron meteorites, which fall less frequently, are the kind most often discovered?

_____

_____

10. Which layer of Earth has a composition that is probably most like the stony meteorites?

_____

11. Why would iron tend to migrate toward Earth's center?

_____

_____

12. What two quantities would you need to know to calculate the average density of Earth?

Evidence about the composition of Earth's core also comes from seismic records. Consider the earthquake epicenter shown in Figure 7-16. Energy waves radiate outward through Earth. From the epicenter to points A and A′, P-waves and S-waves are recorded. However, at station C, on the side of Earth opposite the epicenter, only P-waves arrive. Because S-waves will not penetrate a liquid, scientists infer that the Earth's center contains a liquid.

The zone around Earth between A and B is called the shadow zone. Seismic waves are refracted (bent) at the edge of the outer core. No direct seismic waves (P- or S-waves) can be detected in the shadow zone. The shadow zone encircles Earth like a ring opposite the epicenter, and surrounding the circular region without S-waves.

As the depth within Earth increases, so does the pressure. In fact, the only way to produce such great pressures in a laboratory is to squeeze minerals between the flat faces of diamonds. Tiny bits of iron are squeezed between the diamond faces. The pressure can be increased until one diamond shatters. Meanwhile, x-rays are used to penetrate the molecular structure of the iron. In this way,

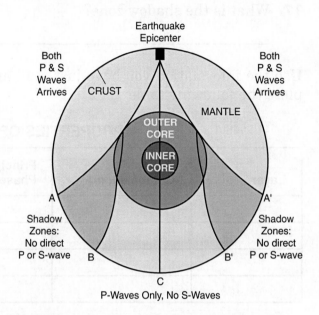

FIGURE 7-16. Seismic waves probe Earth's interior

geophysicists get a good idea of the nature of the inner core. In fact, the pressure at the very center of Earth is high enough to force the molten iron back into the solid state. This occurs in spite of the high temperature at Earth's center. Scientists therefore think that the liquid iron outer core surrounds a solid iron inner core.

Diamonds yield another clue to conditions deep within Earth. To make a diamond, carbon (graphite) must be subjected to extreme heat and pressure. Natural diamonds are found mainly where volcanic explosions have suddenly brought material from very deep within Earth to the surface. Therefore, heat and pressure deep in Earth must be high enough to make diamonds.

Scientists who study geology, astronomy, physics, and chemistry have investigated Earth's interior. As the results of these investigations were shared, scientists constructed a consistent model of the conditions and composition of the interior of our planet.

13. Just as scientists want to know about the inner Earth, sometimes doctors need to know about what is happening inside your body. What medical procedures allow doctors to "see" inside you without surgery? (In fact, there are several.)

What data allows geologists to "see" inside Earth.

14. Which seismic waves cannot travel through a liquid? _____

15. Why do scientists infer that the outer core is a liquid?

_____

16. Why do most seismic waves bend as they travel through Earth?

17. What is the shadow zone?

_____

Use the information you have just read and the *Earth Science Reference Tables* to complete the following table.

PROPERTIES OF EARTH'S INTERIOR

Layer	Composition	Principal Phase/State	Seismic Waves that Pass	Relative Density

Your teacher will show you rocks and other materials to represent those found deep in Earth. Use them to complete the following table.

COMPOSITION OF EARTH WITH DEPTH

Sample	Approximate Depth	Rock Type	Characteristics/Composition
1			
2			
3			
4			
5			
6			
7			
8			
9			

18. According to Figure 7-17, how deep would you have to go to reach Earth's outer core?

_____

19. What is Earth's average density?

_____

20. Why is Earth's crust not shown in this graph?

_____

_____

21. Why does the density line change so quickly at 2900 km?

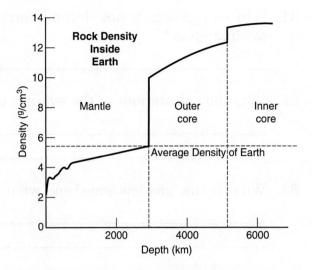

FIGURE 7-17. Earth's density changes with depth below the surface.

_____

Use the Inferred Properties of Earth's Interior chart in *the Earth Science Reference Tables* to answer the following six questions about Earth.

22. What is the average temperature at a depth of 2000 km? _____

23. What is the pressure at a depth of 5000 km?

_____

24. What is the density range of the mantle? _____

25. In what portion of Earth's interior is the temperature consistently higher than the melting temperature?

_____

26. In what part of Earth's interior does the temperature rise the fastest with depth?

_____

27. What is the radius of Earth? _____

28. Why are there question marks near the top of the temperature graph?

_____

29. Why can S-waves not penetrate the outer core? _____

30. Why do scientists study meteorites to investigate the interior of Earth?

_____

31. Why can geologists not obtain information about the deep interior of Earth by drilling holes?

_____

_____

32. What do calculations of the average density of Earth tell us about the interior?

_____

_____

33. What is the "shadow zone" and what causes it?

_____

_____

_____

34. Why are diamonds important to scientists who study Earth's interior? (Hint: There are at least three answers.)

_____

_____

_____

_____

35. Where are kimberlite pipes, rocks that are associated with diamonds, found in New York State? Why do you think there have been no reports of diamonds in these rocks, and why there are no diamond mines or diamond explorations in New York State?

_____

_____

_____

_____

Chapter 8
Plate Tectonics

 CHAPTER 8—LAB 1: GEOPHYSICAL INVESTIGATIONS

Introduction

We know that Earth's crust is dynamic; that is, it is changing. Observations as well as long-term measurements show that the land is being eroded in some places. In other places the land is pushed up to make mountains.

The data given in this lab are fictitious. However, the changes in these quantities are typical of their patterns for planet Earth. These data provide evidence that Earth's crust is dynamic: changing in a paradigm that we know today as plate tectonics.

Objective

In this lab activity, you will use three physical measurements related to surface features on Earth.

PART I: Landforms

Locate and label each of the following geographic features on the part of Figure 8-1 labeled Geographic Features on page 95. Write your labels above the sea-level line.

Mountains	Where the granitic crust reaches its highest elevations
Ocean Ridge	A mostly submerged basaltic mountain range near the center of the ocean
Ocean Trench	The greatest ocean depths, that often occur near a continent
Continental Shelf	The margins of the granitic continents that are covered by shallow ocean water

PART II: Gravitation

The strength of Earth's pull (gravity) at any location on the surface depends on the density of the underlying bedrock. Relatively dense rock, such as basalt, within Earth causes the gravity to be strong at the surface. Gravity is weak where the crust of Earth is composed of rock that is relatively low in density such as granite.

The following data show small changes in the pull of gravity across the region shown at the right of Figure 8-1 on page 000. Positive numbers show where the gravity is slightly stronger than normal and negative numbers show where the gravity is slightly weaker. Plot this data on the section of Figure 8-1 labeled Magnitude of Gravity. Then connect the points with a smooth line.

TABLE 8-1. Variations in the Strength of Gravity

Longitude	Gravity	Longitude	Gravity	Longitude	Gravity
Prime Meridian	+0.1%	35° West	0.0%	70° West	−0.1%
5° West	+0.2%	40° West	0.0%	75° West	−0.2%
10° West	+0.3%	45° West	−0.1%	80° West	0.0%
15° West	+0.2%	50° West	−0.1%	85° West	+0.1%
20° West	+0.1%	55° West	−0.2%	90° West	+0.1%
25° West	+0.1%	60° West	−0.3%	95° West	+0.1%
30° West	+0.1%	65° West	−0.4%	100° West	+0.1%

PART III: Internal Temperature

Earth contains a great deal of heat. Some of it is left over from Earth's formation nearly 5 billion years ago, and some is from the ongoing decay of radioactive elements within Earth. This heat escapes from Earth by heat flow in the form of radiation.

The numbers below the profile of the geosphere shown in the section of Figure 8-1 labeled Heat Flow and Earthquake Foci (Geographic Features) on page 000 represent Celsius temperatures inside Earth. Use these numbers to draw isotherms (isolines that connect places with the same temperature) at an interval of 100°C.

PART IV: Earthquake Depth

Earthquakes happen only where the rocks are brittle. When rock is heated, it becomes more plastic and bends rather than fracturing. Earthquakes are most common in zones of seismic activity. Earthquakes at great depth indicate relatively cool material being drawn into Earth at subduction zones. Rocks can fracture and cause earthquakes only if they are relatively cool and brittle. Plot the 18 seismic events shown in Table 8-2 by drawing a dark X at the proper location on the Figure 8-1 Geographic Features/Heat Flow and Earthquake Foci diagram on page 95.

TABLE 8-2. Locations of Earthquake Foci (Hypocenters).

Longitude (°W)	10	75	78	44	8	69	58	72	96	11	65	12	78	85	71	5	70	66
Depth (km)	4	17	8	3	6	28	0	12	8	5	36	7	11	9	22	8	23	2

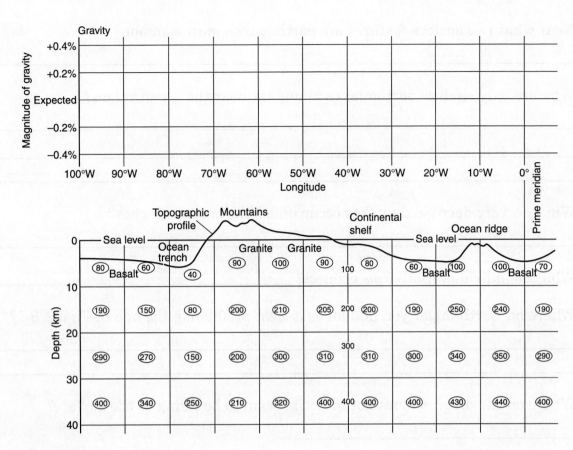

FIGURE 8-1.

Wrap-Up ⟩⟩⟩

1. According to the data you plotted, at what geographic feature is the strength of gravity the highest?

_____

2. What igneous rock is the most common under the oceans? _____

3. Is basalt a low-density or a high-density rock type? (You may use the *Reference Tables.*)

_____

_____

4. What igneous rock is the most abundant within the continents?

_____

5. Why is gravity slightly weaker on the continents than over the oceans?

_____

_____

6. Near what two surface features are earthquakes most common?

_____

7. Why are only shallow earthquakes abundant near the ocean ridges?

_____

_____

_____

8. Why can very deep earthquakes occur under the ocean trenches?

_____

9. What do individual isotherms connect? _____

10. Where on Earth might you find an east-west profile like the one in Figure 8-1?

_____

_____

11. Where on the diagram does hot material seem to be rising to the surface?

_____

12. Where on the diagram is cool material being drawn into Earth?

_____

13. Convection is the circulation of a fluid caused by differences in density. Draw arrows on the Heat Flow and Earthquake Foci section of Figure 8-1 to show convection movements within Earth.

CHAPTER 8—SKILL SHEET 1: PLATE DYNAMICS

The regions where Earth's plates meet form belts of active geologic change, which includes earthquakes, volcanic eruptions, and tectonic mountain building. The relative motion of these plates can be as much as 15 cm/year. However, most plates shift more slowly: about as fast as your fingernails grow. Geologists have identified four types of plate boundaries.

- Divergent rift boundaries occur where two plates pull apart as new crust is created. The mid-ocean ridges are divergent rift zones.
- Convergent subduction zones occur where plates collide and one plate (usually an oceanic plate) dives beneath another. In the oceans, subduction zones are usually found at the deep ocean trenches. As one landmass collides with another, the collision can pile up great mountain ranges.
- At transform boundaries, one plate slides past another.
- Inactive interfaces are places where there seems to be little or no relative motion of the plates.

Actually, most plate boundaries show a mixture of these relative motions: separation, convergence, and slippage. Figure 8-2 is a profile of a diverging rift and a nearby converging subduction zone. Where geologists can locate ancient plate boundaries, they often find associated mineral deposits, such as the copper ores of the American West or emery deposits in Westchester County, New York. (Emery is used as an abrasive.)

On page 9 of the *Earth Science Reference Tables* are diagrams that show the slow motion of the continents over the past 458 million years. Use this diagram to answer the following three questions.

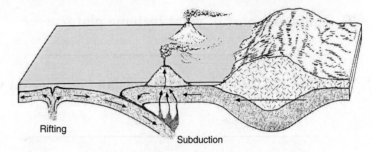

FIGURE 8-2. Rifting creates new crust. Subduction recycles Earth's crust.

1. In what general direction has the North American landmass moved?

   _____

2. The names above the small world maps are geologic time periods. The Atlantic Ocean opened between which two periods?

   _____

3. Approximately how long ago was New York State located at the equator?

   _____

4. What is the most likely composition of crust created at an ocean rift zone?

   _____

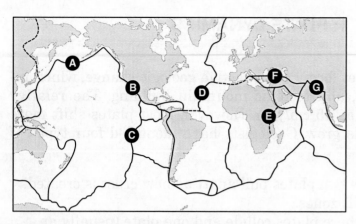

FIGURE 8-3. Earth's major tectonic plates.

To answer the next set of questions you will need to refer to Figure 8-3 and to the Tectonic Plates diagram in the *Earth Science Reference Tables*.

5. Label each of the lettered plate boundaries shown in Figure 8-3 according to the relative motion that is occurring there: Divergent, Convergent, Transform, or Inactive.

A. _____   B. _____

C. _____   D. _____

E. _____   F. _____

G. _____

6. India is a part of what tectonic plate?

_____

7. Where is the only significant land surface on which new crust is being created by rifting?

_____

8. According to this map, why are earthquakes relatively uncommon in New York State?

_____

9. At which kind of plate boundary is part of the crust destroyed?

_____

10. What kind of plate or part of a plate is most like to descend into the interior at subduction zone?

_____

11. Which type of boundary has many earthquakes, but neither creates nor destroys a significant amount of Earth's crust

_____

12. Label the major oceans and continents on Figure 8-3.

CHAPTER 8—LAB 2: HOT SPOTS

Introduction

In plate tectonics, a hot spot is a relatively shallow source of heat below Earth's lithosphere. Figure 8-4 shows the geologic ages of volcanic islands and related features near the center of the Pacific Ocean.

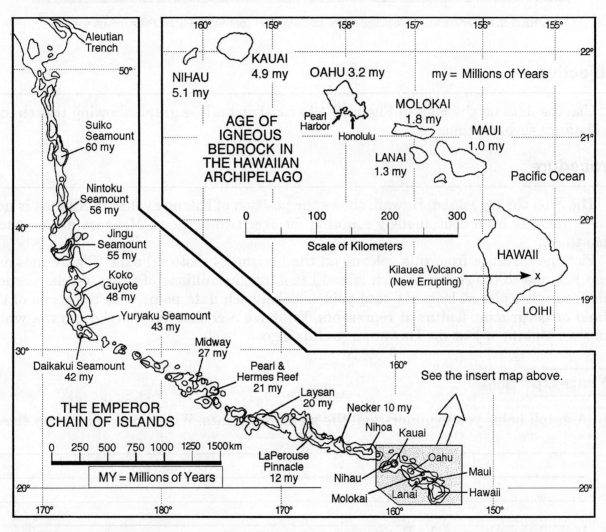

FIGURE 8-4. The Hawaiian Islands are a part of a longer series of islands knows as the Emperor Chain.

From the age data on the map, it appears that the volcanic eruptions have moved southeast through time. However, you are seeing evidence for movement of the Pacific Plate to the northwest, passing over a relatively stationary hot spot in Earth's mantle. As the Pacific Plate moves over the hot spot, upwelling magma pierces the plate forming a series of volcanoes that appears to extend toward the southeast. However, the real motion is the tectonic plate traveling northwest. (See Figure 8-5.)

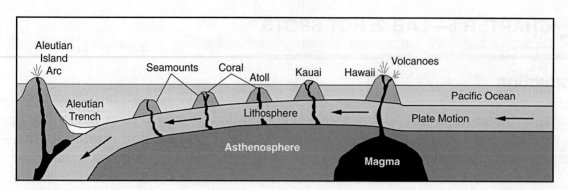

FIGURE 8-5. Volcanic islands are pulled off the stationary hot spot as the Pacific Plate moves northwest.

Objective

Use the data on the map in Figure 8-4 to construct a line graph showing this change in distance through time.

Procedure

The X on the big island, Hawaii, shows the position of Kilauea volcano. Kilauea is now active and has been continuously erupting for over two decades. Measure all distances from the X.

Plot the distance from this volcano on the horizontal (bottom) axis. (Please measure from Kilauea to the center of each island.) Plot age, in millions of years, on the vertical axis. You must plot at least six data points. Label each data point with the name of the island or geographic feature it represents. You have a choice. You may graph the whole Emperor Chain, or just the Hawaiian Archipelago.

Wrap-Up ▶▶▶

1. A graph helps you to understand the meaning of data. What does your graph show?

   _____

   _____

   _____

2. If the lithosphere of the Pacific Ocean is moving over a stationary hot spot within the mantle, what is the average speed of this motion in cm/year? (Remember that 1 km = 100,000 cm.) To change kilometers to centimeters, just add five zeros. (Please show your calculation clearly, and round your answer to the nearest whole number with the proper units. Significant figures might help.)

3. Where is the next volcanic island likely to appear?

_____

4. These islands are forming on top of the moving Pacific Plate. In what direction is the Pacific Plate moving?

_____

5. The slope of the line on your graph shows the rate of plate movement. How could you tell, from a quick look at your graph, whether the speed of movement was constant?

_____

6. When does it appear that the Pacific plate changes its direction of movement relatively suddenly?

_____

7. What is a geologic "hot spot"? _____

_____

8. What are the terrestrial coordinates of the city of Honolulu? (Please express your answer to the nearest whole degree.)

_____

9. Compare the rate of motion of the Hawaiian Islands with the rate of motion of the whole Emperor Chain. Has the motion accelerated or slowed through time? Justify your answer.

_____

_____

_____

_____

_____

My Notes

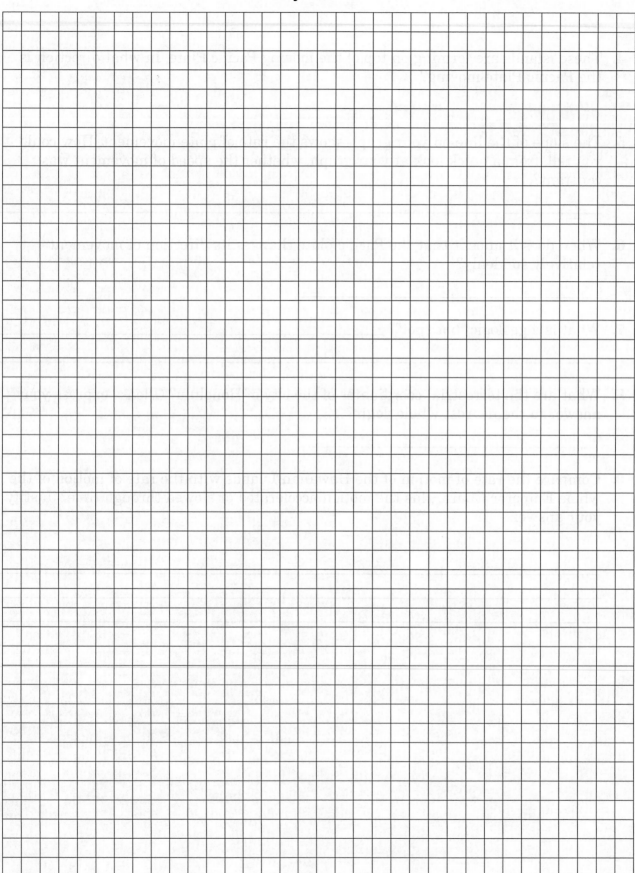

Chapter 9
Geologic Hazards

 CHAPTER 9—SKILL SHEET 1: GEOLOGIC HAZARDS

Seismologists sometimes say, "Earthquakes don't kill people. Buildings do!" In most earthquakes, the greatest loss of life and property happen in the collapse of human-made structures. Most buildings in the United States are built on a wood or steel frame. However, in the past and in today's less developed nations, many people live in homes made of adobe, stone, or other materials, which may resist damage from vertical (up-and-down) motions. However, these buildings have no internal framework. The primary importance of the framework is to resist horizontal (side-to-side) forces and movement.

The following passage illustrates the importance of safe construction. An earthquake in China, which occurred in 1556, probably caused the greatest loss of life in a seismic event. This catastrophe is estimated to have killed nearly a million people. In 2005, a magnitude 6.4 earthquake in central Iran killed more than 600 people. The buildings in this area were not built to withstand earthquakes. However, a slightly larger quake in California 2 years earlier killed only two people. In California, buildings must be constructed so they are resistant to damage from earthquakes.

Damage is especially severe where structures have been built on thick sediments. Low-density sediments vibrate more violently than solid rock. Furthermore, sediments do not provide as secure a foundation as bedrock. In addition, water-saturated sediments are subject to liquefaction. When this occurs, solid ground becomes like quicksand, causing structures to tip or partially sink. In addition, thick sediments actually make the shaking at the surface more violent. The 1989 Loma Prieta, California, earthquake was caused by fault slippage in the Santa Cruz Mountains. However, most property damage and loss of life occurred in the San Francisco Bay area about 80 km (50 miles) away. This was because the San Francisco Bay Area has a high population density and is largely built on loose sediments. (See Figure 9-1.)

In addition to the collapse of structures, there are other earthquake hazards. A massive landslide triggered by

FIGURE 9-1. Earthquake damage in San Francisco, 1989.

an earthquake in Wyoming in 1959 buried dozens of people. The San Francisco earthquake of 1906 led to a fire that destroyed most of the city. Broken water mains and rubble in the streets made fighting the fire impossible. A tsunami is a series ocean waves that can be generated by an earthquake. These waves often increase in height as they move into shallow water. Shortly after the 2004 Indonesian earthquake, approximately 300,000 people were killed by the tsunamis. Large portions of coastal cities along the Indian Ocean were devastated by waves as high as 10 meters (35 feet).

1. In the movies, you sometimes see people fall into a large crack that opens to swallow its victims. Then the opening closes like the jaws of a crocodile. However, it is far more likely that the unstable sides of the pit simply collapsed. What events associated with earthquakes cause the greatest loss of life and property?

2. What is a tsunami? _____

3. If you plan to build in an area with frequent earthquakes, choosing what kind of location would reduce the shaking and the chance of building collapse?

4. Name three earthquake-related events that can cause damage and loss of life after the ground stops shaking.

    a. _____

    b. _____

    c. _____

How Can You Prepare for an Earthquake?

5. Make a list of the potential hazards in and around your school building.

6. Devise a practical plan to reduce or eliminate these problems. Tell what must be done and estimate the cost of each modification. Try to keep costs down by suggesting simple and inexpensive changes whenever possible.

_____

_____

_____

_____

7. Write a plan of action for teachers and students to follow in the event of an earthquake. Keep in mind that some earthquakes are far more damaging than others.

_____

_____

_____

_____

How Do You Deal with Other Geologic Hazards?

Make a list of two to three other geologic hazards that may not be associated with earthquakes. For each event, tell how to reduce or eliminate the danger to people and property. (See Figure 9-2.)

FIGURE 9-2. Mount St. Helens in Washington.

Chapter 10
Weathering and Soils

CHAPTER 10—SKILL SHEET 1: WEATHERING

Weathering is the adjustment of Earth materials to conditions at or near Earth's surface. Many rocks that are stable inside Earth begin to change when they are exposed to the atmosphere. There are two types of weathering. **Physical weathering** occurs when a rock breaks into smaller pieces or is dissolved by water. (Change in state is also a physical change.) If the material changes into a new substance, **chemical weathering** has occurred.

1. What is weathering? _____

   _____

2. How do chemical changes differ from physical changes?

   _____

   _____

Figure 10-1 shows which type of weathering is dominant under different conditions of climate. A major form of weathering in cold, moist climates is frost action. When water within cracks in rocks changes to ice it expands. This forces the cracks open as the temperature cycles above and below freezing. In moist and warm climates, water causes many forms of chemical weathering, such as oxidation (rusting), to accelerate.

3. How does freezing water cause physical weathering?

   _____

   _____

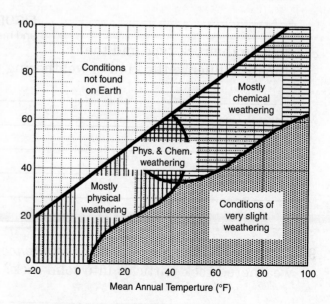

FIGURE 10-1. Weathering is controlled by rainfall and temperature.

4. In what type of climate does chemical weathering dominate?

_____

5. Draw a dark X on Figure 10-1 to represent your local climate.

_____

6. Based on Figure 10-1, what can you say about the weathering that most likely occurs in you area?

_____

The dominant form of weathering also depends on the type of rocks found in a particular area. The minerals in sedimentary rocks are relatively stable under conditions at Earth's surface. Therefore, sedimentary rocks tend to be weathered by physical processes. Igneous and metamorphic rocks are usually made from intergrown crystals that formed deep underground. Exposed at the surface, these rocks are more likely to break down by chemical weathering.

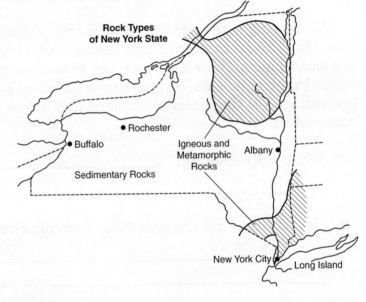

7. Draw an X on the map at your geographic location. According to this map, which group of rocks occurs in the area where you live?

_____

_____

_____

FIGURE 10-2. The regions of New York State where igneous and metamorphic rocks dominate.

8. How do the rock type and the climate influence the kind of weathering that is dominant in your location?

_____

_____

_____

_____

_____

9. According to the *Earth Science Reference Tables,* what processes can change weathered rock particles into solid rock?

_____

_____

CHAPTER 10—LAB 1: ROCK ABRASION

Introduction

As rocks are tossed about in a fast flowing stream, they constantly collide with each other, with objects floating in the stream, and with the stream's banks and bed. The friction generated by the collisions wears away the rocks by abrasion. Abrasion also occurs at a beach, where waves throw sand or other particles on the beach. It also occurs where sand particles blown by the wind blast other rocks. Abrasion also occurs as rocks grind together during transport by glacier. Abrasion is an important form of physical weathering.

Objective

In this lab you will measure the ability of different kinds of rock to resist weathering by abrasion. You will use your data as well as data from other lab groups to complete this lab.

Materials

Each Group: Mass scale, wide-mouth plastic bottle that closes tightly
For the Class: Plastic buckets with different kinds of rock chips soaking in water, buckets used only for rock remains, strainers (to stay with the buckets), clock or timer, paper towels, sponges

Procedure

1. Your teacher will assign your group a type of rock. What kind of rock has your group has been assigned to abrade?

2. From the large container of soaking rocks, take a small handful of wet rocks. Use the scale to measure 100 grams of wet rock (± 0.5 gram). Return the rock you do not use to the container of soaking rocks.

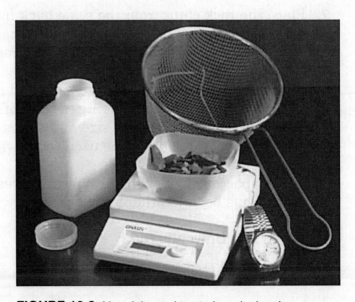

FIGURE 10-3. Materials used to study rock abrasion.

3. Add about 2–3 cm of water into a wide-mouth plastic bottle. Screw the top on tightly and shake the bottle to check for leaks. If the bottle leaks, tell your teacher. If the bottle does not leak, add your 100 grams of wet rock to the water in the bottle.

4. Secure the top tightly on the bottle, and shake the contents for four (4) minutes. It is important that you shake the rocks vigorously for the whole time. In addition, the energy of shaking should be constant.

(In the next step, do not allow rock chips to go directly into the bucket.)

5. Take your bottle to the used rock bucket. Dump your rocks into a strainer held inside the bucket of used rocks. Wash the rocks in the strainer and the inside of the bottle, with water from the bucket. Then drain the rock fragments, and tip them onto a dry paper towel. Leave the strainer in the bucket for the next group.

6. Measure the mass of rock remaining to the nearest 0.1 gram and record your data in Table 10-1.

TABLE 10-1. Rock Abrasion Data

Rock Type	Members of Lab Group	Initial Mass	Mass after 4 Minutes of Shaking	Mass after a Total of 8 Minutes of Shaking	Mass after a Total of 12 Minutes of Shaking
Rock Salt					

7. Place the rock chips from the scale back into the bottle with about the same amount of water. Shake it for an additional 4 minutes (total of 8 minutes). Then wash, drain, and record the mass of the remaining rock fragments.

8. Repeat this procedure one more time and record the remaining mass after an additional 4 minutes of shaking (a total of 12 minutes of shaking). Note how the rock chips have changed in size and shape. Then dump the rocks into the "used rock" bucket. Do not mix the used rocks with fresh rocks.

9. In the spaces below, sketch a typical unweathered rock chip and the same chip after it has been weathered by abrasion.

Unweathered

Weathered

10. Repeat the procedure with rock salt. (Your teacher can help you with this step.)

11. Obtain data collected by other lab groups to complete Table 10-1. You will be responsible for obtaining reasonable data. If the data does not seem reasonable, you can either find better data from a different group, or you can do the procedure yourself. You must have data for four rock types.

12. Graph your data. Put Time on the horizontal axis and Mass remaining on the vertical axis. You will draw four lines (one for each kind of rock) on the same graph. Your graph will show how quickly each of the four kinds of rock was abraded.

13. Which two rock types seem to be most similar in their resistance to abrasion? Why are they similar?

Wrap-Up

1. Define abrasion. Your definition must include how abrasion works.

_____

_____

2. Is abrasion a form of weathering (breakdown), or erosion (transportation)?

_____

3. Is abrasion a physical or a chemical change?_____

4. In nature, where are rocks abraded by this process? (Please be specific.)

_____

5. Of the rocks in your data table, which rock type was abraded the most?

_____

6. What rock type used in this experiment is very susceptible to chemical weathering, but resistant to physical weathering?

_____

7. Why did the rocks lose more mass in the first 4 minutes of shaking than in the following 4-minute intervals of shaking time?

_____

8. What two properties of rock salt caused it to wear away so quickly?

     a. _____   *b.* _____

9. What fraction of your first rock was left after a total of 12 minutes of abrasion?
Please circle the closest answer.

$\frac{19}{20}$ $\frac{9}{10}$ $\frac{17}{20}$ $\frac{4}{5}$ $\frac{7}{10}$ $\frac{3}{5}$ $\frac{1}{2}$ $\frac{1}{3}$ or $\frac{1}{10}$

10. How did the rock fragments change in size and in shape as they were abraded?

_____

11. Figure 10-4 shows two rocks of similar composition. In what two ways does the rock on the right differ from the rock on the left? Why do they differ?

FIGURE 10-4. Two samples of diorite.

_____
_____
_____
_____
_____
_____
_____

12. What plumbing feature built under most sinks keeps rock fragments and other dense sediment from settling into the underground plumbing, where it is difficult and expensive to clean out? (You may name it or you may describe it.)

_____
_____

CHAPTER 10—LAB 2: INQUIRY INTO SURFACE AREA

Introduction

Weathering, like most changes, happens primarily at interfaces. An interface is a boundary between different materials or different systems. Figure 10-5 and the diagrams below show that slicing a sample into smaller pieces increases its surface area. The rate at which any material reacts in a chemical change depends on the total amount of surface area exposed. Weathering breaks rocks into smaller pieces. This creates additional surface area.

This activity may be a different from the labs you have done before. In fact, you will be designing and conducting your own laboratory procedure. In this lab, the title, objective, and materials have been provided. You must devise and follow your own procedure to collect data, make a graph of the data, and draw as many scientific and mathematical conclusions as your observations support. Students must create and record the Procedures, Data, and Conclusions.

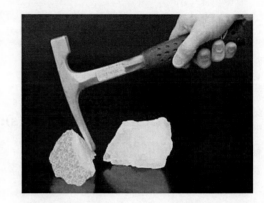

FIGURE 10-5.

Guidelines

- You should submit one lab paper for the group (maximum 3 people).
- If possible, reports should be printed on a word processor.
- Write the six parts of the lab in proper order. (The graph can be on a separate sheet.)
- You will need at least three data points to make your graph line.
- Be sure your conclusion(s) fits the objective.
- Make it easy for yourself. Keep you lab report brief. For example, the procedure need not specify details that a reasonable person would not need.
- Your procedures must conform to laboratory safety rules.

Title: Surface Area

Objective: To investigate how the surface area of a cube changes as the cube is divided into smaller pieces, and to construct a graph to that illustrates that change.

Materials: Graph paper, a large cube that can be disassembled into smaller cubes and other materials as requested.

FIGURE 10-6.

Procedure:

Observations/Data/Graph(s):

Conclusions:

CHAPTER 10—LAB 3: INQUIRY INTO TEMPERATURE AND WEATHERING

Introduction

For this lab, you again will be designing and conducting your own laboratory procedure. Use the following open-ended format:

Title: Temperature and Weathering

Objective: To investigate how temperature affects the rate of chemical weathering.

Materials: Several pieces of Alka-Seltzer™ or a similar effervescent tablet per group, other available materials and lab ware as requested

Procedure:

Observations/Data/Graph(s):

Conclusions/Discussion:

In this lab, the title, objective, and materials have been provided. You must devise and follow your own procedure to collect data, make a graph the data, and draw as many scientific and mathematical conclusions as your observations support.

Students must create and record the Procedures, Data, and Conclusions.

FIGURE 10-7. Collaboration is an important part of science.

Guidelines

- You should submit one lab paper for the group (maximum 3 people).
- If possible, reports should be printed on a word processor.
- Write the six parts of the lab in proper order. (The graph can be on a separate sheet.)
- You will need at least three data points to make your graph line.
- Be sure your conclusion(s) fits the objective.
- Make it easy for yourself. Keep your lab report brief. For example, the procedure need not specify details that a reasonable person would not need.
- Your procedures must conform to laboratory safety rules.

My Notes

Chapter 11
Erosion and Deposition

 CHAPTER 11—LAB 1: SETTLING RATES

Introduction

Rivers and other streams slow when they empty into a lake or other large body of water. As the current becomes slower and less turbulent, particles settle out of the calm water. The settling out of sediments is known as deposition.

Objective

To determine the factors that affect the rate at which sediments are deposited.

Materials

Plastic column (capped at the lower end, resting in a large plastic container), ring stand or other support, stopwatch or timer, masking tape, ruler or meterstick, 3 glass marbles (about 12 mm), steel marble, big lead sphere (about 12 mm), 2 spherical stones (1 large, about 12 mm, and 1 much smaller stone), glass disk, tiny lead shot, 2 plastic beads (1 large and 1 small), tiny glass bead

Procedure

STEP 1: Gather Materials
Obtain a set of settling rate objects from your teacher. Then arrange them on your desk. You need to do this for two reasons. First, you will need a full set to complete the lab. Second, you must learn how these objects match with their names, so you know which ones to use in each step of the procedure.

STEP 2: Density and Settling Rate
(Glass marble, steel sphere, lead sphere, and large plastic sphere—all approximately the same diameter)

FIGURE 11-1. Finding settling times.

Be sure that the plastic cap is securely attached to the bottom of the column and that it does not leak. Fill the column with water to within about 3 to 4 cm of the top (as shown in Figure 11-1). Rest the tube in a plastic container. Use masking tape to mark two positions on the tube: one near the top and a second near the bottom. Each settling time should be measured between higher mark and the lower mark.

Arrange the three large spherical objects (plastic, glass, and lead) on your desk in order of density from the least dense to the most dense. Drop the objects one at a time as shown in Figure 11-1. Use the stopwatch or timer to see how long it takes each object to travel from the higher mark to the lower mark on the column. Record in Table 11-1 the settling time for each of the three particles. Do not forget to specify the units of measure.

TABLE 11-1

Density	Object	Settling Time (with units)	Settling Rate (with units)
Least Dense			
Most Dense			

1. Which of these particles settled fastest? Why?

_____

2. As the density of the particle increases, what happens to the settling time?

_____

3. As the density of the particle increases, how does the rate of settling change?

_____

4. How are settling rate and settling time related? _____

_____

5. *a.* If they were released at the same time, predict which object will settle first: the small lead shot or the plastic bead of the same size?

_____

** *b.*** Why did you pick that one? _____

** *c.*** Try it. Was your prediction correct? _____

6. As the rate of settling increases, the time to settle _____

STEP 3: Size and Settling Rate
(Small pebble, large pebble, second glass marble, tiny glass bead)

7. Drop the small stone and the large stone together into the column of water.

Which settles more quickly? _____

8. Next, drop the marble and the tiny glass bead together.

Which of these glass objects is first to be deposited? _____

9. If the shape and density are equal, what effect does the particle size have on the rate of deposition?

_____

_____

10. As particle size increases, the rate of settling _____

11. As the size of the particles increases, the settling time _____

STEP 4: Shape and Settling Rate
(Third glass marble, glass disk)

Drop the glass disk into the column, use a stopwatch or timer to record how long it takes the glass disk to settle to the bottom of the column. Do the same with the glass marble. Answers will vary, but sould include the appropriate unit.

12. *a.* Settling Rate: Disk _____ Marble _____

** *b.*** Which shape settles faster? _____

** *c.*** Why? _____

Repeat this experiment with a different fluid, air. Take two sheets of scrap paper. Crumple one of them into a tight ball and leave the other sheet open. Then hold them high above the floor and drop them.

13. Which reached the floor first and why? _____

_____

14. What is a fluid? _____

STEP 5: Clean Up

Carefully and slowly empty the column of water with all the objects into the plastic container. (The water tends to rush out suddenly.) Please check the set of objects to be sure that the next group will have a complete set, with no extra objects or no objects missing. (Do not take objects from a different set.) Your teacher can supply missing objects or take any extras.

Wrap-Up ▶

15. Define deposition. _____

16. How does the *density* of particles affect the rate at which they are deposited?

_____

17. How does the *size* of particles affect how quickly they are deposited?

_____

18. How does the *shape* of particles affect the rate of deposition?

_____

19. What three characteristics of density, size, and shape cause particles of sediment to settle quickly in water?

_____

20. Sketch the expected relationship between particle density and settling time and particle density and settling rate.

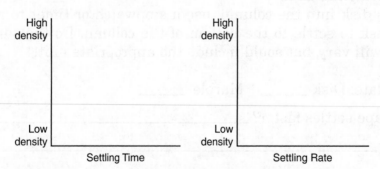

FIGURE 11-2. Graphs to sketch settling time and settling rate.

21. Compare and contrast settling time and settling rate.

_____

_____

CHAPTER 11—LAB 2: DEPOSITION IN WATER

Introduction

When a stream carrying sediment slows, it deposits its sediment.

Materials

Part I: Well-sorted beach sand, beaker, teaspoon, 400-mL beaker
Part II: Tall plastic column, water, poorly sorted sand

Objective

To study the factors that affect deposition in water.

Procedure

Part I: Erosion and Deposition at a Meander
Put about ½ teaspoon of well-sorted beach sand into a 400-mL beaker. Then fill the beaker about half way with tap water. Slide Figure 11-6, on page 124, under the beaker, centering the beaker on the circle.

Stir the water with the spoon until it swirls steadily as shown in Figure 11-3. Then take the spoon out and watch the sand as it settles to the bottom of the beaker.

1. Where does most of the sand settle in a stream bend?

   _____

   _____

2. Why does it settle in that place?

   _____

FIGURE 11-3. Making a model of a curve in a stream.

3. What does this procedure tell us about where erosion and deposition will occur in a stream meander? (A meander is a bend or curve in a stream.)

   _____

   _____

Rinse your beaker upside down in the wet sand bucket to clean it. Please keep sand out of the sink and drain.

Part II: Deposition of Graded Bedding
Set up a tall plastic column filled with water to within about 15 cm of the top. Use a dry funnel to dump very quickly about 40 mL of poorly sorted sand into the column. Watch the sand settle.

4. Which particles settle at the fastest rate? (List three characteristics.)

_____

Wait a minute or two for the smaller particles to settle. The layer you just made is called a layer of graded bedding. Notice where the particle size changes gradually, and where it changes suddenly at a sharp interface. (If you do not see these sudden and gradual changes, ask your teacher to help you.)

5. Is the change in particle size within a single layer a sharp change or a gradual change?

_____

6. Describe the one layer of graded bedding that was deposited by one beaker of sand.

_____

_____

7. In Figure 11-4, carefully draw exactly two layers of graded bedding. Be sure to show clearly where the particle size changes gradually and where it changes suddenly.

8. How might graded bedding form in a natural setting?

_____

_____

FIGURE 11-4.

Separation of particles by size is called sorting. The sand you dumped into the tall column was poorly sorted sand, although it became partly sorted by the time it settled. Sediments deposited by wind and water are usually better sorted than sediments deposited by glaciers or gravity falls.

As a stream enters the calm water of a lake or ocean, the larger particles are deposited first. The smaller particles of sediment are carried farther from shore. Figure 11-5 shows how a stream entering calm water leaves deposits with horizontal sorting.

9. According to the *Earth Science Reference Tables,* why are the smallest particles carried farthest from the shore?

_____

_____

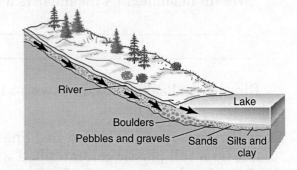

FIGURE 11-5. As a stream reaches calm water, the largest particles are deposited first.

10. How are ions (individual charged atoms), which may never settle, carried by water?

_____

Complete Table 11-1 based on information in the *Earth Science Reference Tables.*

TABLE 11-1. Sediment Texture and Size Range

Sediment Name	Texture	Size Range	Minimum Stream Velocity Needed To Move This Size Particle (cm/sec)
Clay	Tiny particles that feel silky	Less than 0.0004 cm	Near 0 (less than 10)
	Grains barely visible with a hand lens	0.0004–0.006 cm	Approximately 10
	Particles feel gritty		Approximately 20
	Composed of small stones		Approximately 50
	Rocks the size of a fist		Approximately 170
	Rocks larger than your head	Larger than 25.6 cm	Approximately 280

11. According to the *Earth Science Reference Tables,* which particle size needs the least energy to move it? (Write the name.)

_____

12. How fast must the current be to move an average size cobble?

_____

13. What is the largest diameter of common rock sediment that can be moved by a stream current moving at 2.5 meters per second?

_____

14. Which sedimentary particles are larger than clay, but smaller than sand?

_____

15. Define deposition. _____

16. What do we mean by the sorting of sediments?

_____

17. What two agents of erosion and deposition sort sediments by particle size?

_____

18. On Figure 11-6, write Erosion where erosion occurs, and write Deposition where deposition is likely.

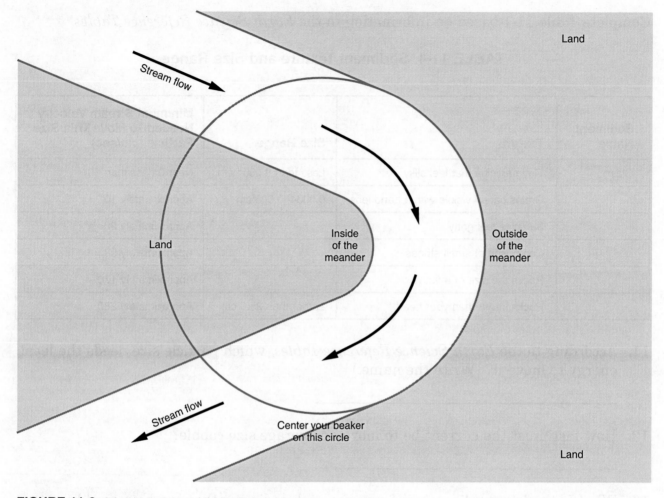

FIGURE 11-6. A beaker as a model of a stream.

Chapter 12
River Systems

 CHAPTER 12—LAB 1: STREAM VELOCITY

Introduction

Hydrologists are scientists who study the circulation of water in and on Earth. They often use models to represent real Earth systems. Figure 12-1 is a picture of the equipment used to model a stream. With this model we can change variables that would be difficult to change in a real stream. By changing these factors, we can investigate the causes of changes in the velocities of real streams.

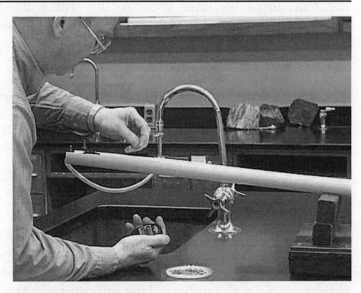

FIGURE 12-1. A model of a natural stream.

TABLE 12-1. Stream Velocity Data

Stream Velocity	Shallow Gradient		Steep Gradient	
	Time	Rate	Time	Rate
Low Discharge				
High Discharge				

1. What is the length of your stream model? _____

2. How does the gradient of a stream affect the velocity of the stream?

_____

_____

3. As the volume of water in a stream increases, what happens to the velocity of the water? Why?

_____

_____

Water flows downhill pulled by the force of gravity. Friction between the moving water and the banks and bed of a stream (and even the air) slows the current as shown in Figure 12-2. The speed of any stream is determined by a balance between the forces of gravity and friction. If gravitational force becomes greater, the water flows faster. If friction dominates, the stream will slow.

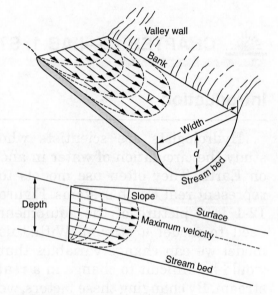

FIGURE 12-2. Stream velocity is greatest where the influence of friction is least.

4. What force slows stream water?

_____

5. According to Figure 12-2, where in a stream is the current fastest? Why?

_____

_____

Bends in a stream are called meanders. At a meander, inertia swings the fastest flowing water to the outside of the bend. Inertia is the property of matter that resists changes in motion. It causes your body to continue moving in a straight line when the car in which you are riding makes a sharp turn. This difference in the water's speed causes erosion on the outside of the stream bend, and deposition along the inside. This is shown in Figure 12-3. The

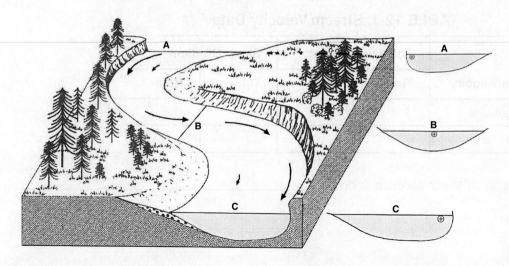

FIGURE 12-3. Streams flow fastest at the outside of a meander where water is deepest.

difference in velocity also causes the stream to be deeper near the outside of a meander, and shallower along the inside, where the water flows more slowly.

6. What is a meander? _____

7. Why is the current fastest near the outside of a meander?

_____

8. Erosion usually dominates along the **a.** _____ of a meander, while along the inside of the same meander **b.** _____ generally dominates.

Continuing erosion and deposition cause meanders to change and to move slowly downstream through time. Figure 12-4 shows the changing path of a section of the lower Mississippi River over a period of 160 years. Moss Island, which formerly connected to Louisiana, is now on the Mississippi side of the river. Where a political boundary follows a meandering river, confusion may result as the river changes its course through time.

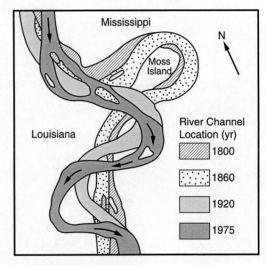

FIGURE 12-4. Erosion and deposition cause meandering streams to change through time, as with the meandering Mississippi.

9. What two processes change the course of meandering rivers?

a. _____

b. _____

Changes in the discharge of a stream cause changes in the speed of the water. Sediment deposited in the stream channel when the water's velocity slowed may be washed away in a flood as shown in Figure 12-5. Notice how, month-by-month, sediment builds up in the streambed until it is washed downstream by a flood.

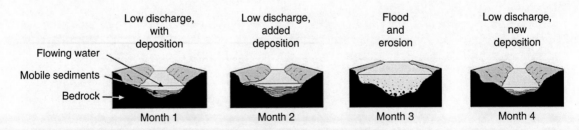

FIGURE 12-5. Deposition and erosion are a function of water velocity.

When the continental glaciers that covered most of New York State melted back, about 20,000 years ago, flooding scoured the channel of the Hudson River. More recently, thick layers of sediments have been deposited along most of the river bottom.

10. What would it take to move the sediments on the bottom of the Hudson River?

_____

11. When people look at our largest rivers, such as the Mississippi River near New Orleans, the calm flow leads them to think that most rivers slow as they flow toward their mouth. After all, the stream gradient decreases dramatically. However, measurements show that streams generally flow faster as they move downstream, even as the slope decreases. Why do rivers generally flow faster as they flow downstream?

_____

CHAPTER 12—SKILL SHEET 1: HUDSON RIVER PROFILE

The Hudson River is sometimes called "New York's Main Street" because of its historical use as a major transportation corridor. Canals, roads, and railroads followed this gateway to the American interior. Use the data below and a sheet of graph paper to draw a longitudinal profile of the Hudson River. The Hudson flows from its source in the Adirondack Mountains to its mouth in New York Harbor. The points of interest do not need to be shown on your graph.

Distance (miles)	Elevation (feet)	Geographic Points of Interest
0	4293	Lake Tear of the Clouds, near the summit of Mount Marcy, the source of the Hudson River on New York's highest mountain.
1	3564	Feldspar Brook, the outlet to Lake Tear of the Clouds, flows into the Opalescent River
3	2764	Lake Colden Ranger Station has an excellent view of New York's high peaks
9	1755	Henderson Lake's outlet is called the Hudson River.
59	1028	Town of North Creek, near Gore Mountain ski area, holds white-water canoe races on the Hudson River each May.
109	294	Located in the town of Hudson Falls, General Electric was fined $7 million for dumping PCBs into the river. The PCBs contaminated game fish in the river.
152	14	Ocean tides reach all the way to Federal Dam at Troy. The Mohawk River, the Hudson's largest tributary, enters near this point.
158	2	A channel with a minimum depth of 32 feet is maintained as far north as the Port of Albany, the capital of New York State.
229	1	Below the city of Poughkeepsie, mixing of the river caused by the tides makes the Hudson River increasingly salty.
249	0.5	The deepest point in the river, 202 feet below sea level, is at West Point Military Academy.
276	0.1	Tappan Zee, 3.2 miles across, is the widest part of the river. Zee is the Dutch word for sea.
304	Sea level	At the mouth of the Hudson River, Upper New York Bay is one of the world's busiest harbors. The average discharge of the river is about 20,000 cubic feet per second.

The concave shape of the Hudson River profile is typical of large streams. However, perhaps the most unusual feature of the Hudson River is that the river is a sea level tidal estuary (arm of the sea) for nearly half of its length.

1. Why has the Hudson River valley been important to the development of the interior?

_____

_____

_____

2. How does the gradient of the Hudson River change with distance from its source?

_____

_____

3. Vertical exaggeration = unit distance on the horizontal scale divided by the value of the same distance on the vertical scale. Note that the units of both must be the same. What is the vertical exaggeration of your profile?

_____

_____

4. The Hudson River flows directly through New York City. Even if it were not polluted, why would the city not use the Hudson as a source of drinking water?

_____

_____

_____

CHAPTER 12—SKILL SHEET 2: WATERSHEDS

A watershed is the geographic region in which water drains into a particular stream or river. It is sometimes called a drainage basin. Watershed maps are useful if you need to understand where water comes from and to predict the flow of waterborne pollutants.

Figure 12-6 is a map showing the major streams in New York State. Use the *Earth Science Reference Tables* to help you label the major rivers of New York State. Then use this map to outline the Hudson River watershed. For example, the drainage basin of the Genesee River in western New York and northern Pennsylvania has been outlined with a dashed line. This dashed line separates the streams that flow into the Genesee from those that do not flow into this watershed.

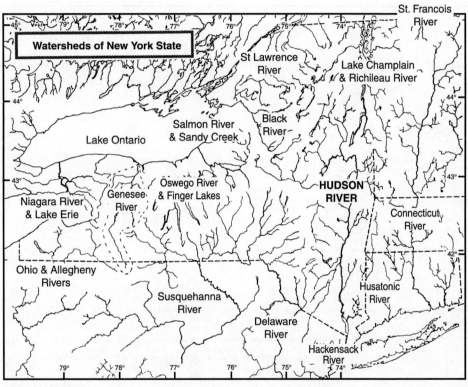

FIGURE 12-6.

1. Draw an X on the map to show where you live. In what watershed do you live?

_____

2. What is the largest river that is totally within New York State?

_____

_____

Name Teacher Period Date

3. It is easy to see where water flows in streams and rivers. However, water in the ground is not visible. In what general direction does most groundwater flow in New York State?

4. What land features are generally at the boundaries of watersheds?

5. Why is it important for an informed citizen to understand the extent of her/his local watershed?

Chapter 13
Groundwater

 CHAPTER 13—LAB 1: WHY IS WATER WEIRD?

Introduction

Water has many remarkable properties. For example, unlike most substances, water expands when it becomes a solid. Water is the only common substance that exists on the surface of our planet in all three states: solid, liquid, and gas. It is also one of the most efficient absorbers of heat energy. That is why immersion in water is one of the best ways to cool hot metal very quickly. Given enough time and quantity, water can dissolve more substances than any other natural liquid. Pure water has a strong surface tension, which pulls a drop of water into the smallest possible volume. In addition, surface tension lifts

FIGURE 13-1.

water hundreds of feet to the tops of the tallest trees. Although surface tension may seem like a very small force, water's surface tension can be remarkably strong.

Materials

(Please pick them up only as you need them.) Dropper pipette, metal or plastic cup, metric ruler, paper towel or tissue, penny, straight pins, tweezers, liquid detergent, sponge

Objective

To learn about some unusual properties of water.

Procedure

1. Clean a new penny with a tissue or a paper towel. Then place it on a flat surface. Use a dropper pipette to deposit drops of water on the penny. How many drops of water can you balance on top of the penny before it spills?

2. *a.* How does the penny look when you view it through the crown of water?

   b. Why does this occur?

3. Place the cup where a water spill is not likely to be a problem. Fill the cup until the water is exactly level with the rim of the cup. (You will need to observe this from the side.) Carefully add as much water as you can (perhaps with the dropper) until the water starts to spill over the edge of the cup. As accurately as you can, measure how high (in millimeters) the crown of water extends above the rim of the cup.

4. Use the tweezers to place as many pins as you can on top of the water. Each pin must be placed slowly so it rests flat on the water surface. How many pins can you "float" before they all sink?

5. *a.* Why do steel pins "float" on top of the water?

   b. Why do ships float?

   c. Are ships also held up by surface tension?

6. Why do the pins move to the side of the cup?

   Now pour out the water and pins. Take care to recover all the pins.

7. Fill the cup with water again. Float a single pin on the surface. Add a tiny drop of detergent. What happens?

_____

8. Can you place pins on the water surface after you have added the detergent?

_____

Why? _____

_____

Return all the pins and other lab materials.

Wrap-Up

9. What property of water supports gliding insects such as water striders?

_____

10. How much more dense are the steel pins than water? (You can find the density of steel experimentally, in a book, or on the Internet.)

_____

11. Water is sometimes used as a standard.

    a. What is the density of water in metric units? _____

    b. On the Celsius scale, what is water's freezing point _____ and boiling point _____? (These temperatures change with major changes in air pressure.)

12. If you feel hot on a summer day, why is water so useful in cooling you off? (There are several answers.)

_____

_____

_____

My Notes

⊕ CHAPTER 13—LAB 2: POROSITY

Introduction

What properties of a soil make it a good aquifer? One of them is porosity. The porosity of a material is a measure of the portion of open space between the soil grains. Some soils are more porous than others. Clay usually has a low porosity because the particles often are flat and they pack closely together. Sandy soils tend to be composed of rounded particles that have been sorted by wind or water into particles of the same size. Rounded grains cannot pack closely, so sandy soils often have a high porosity. When a rock is cracked or broken up, its porosity usually increases. Porous rocks and soils are better at holding water, and they usually drain more readily than compact soils.

FIGURE 13-2.

Objective

To determine the porosity of a sample of sand.

Materials

Small graduated beaker (about 150 mL), 100-mL graduated cylinder (dry), well-sorted beach sand

Procedure

1. Select a dry graduated beaker. If it is not completely dry, wipe it dry or select another beaker.

2. Put 100 mL of dry sand into your beaker.

3. Measure 100 mL of water with a graduated cylinder.

4. Very slowly pour the water into the sand until the water's surface is even with the top of the sand. (If you add too much water, you can carefully pour some back into the graduated cylinder.)

5. How much water is left in the graduated cylinder (including any you poured back)?

Therefore, how much water was used to fill the spaces between the sand grains?

6. Calculate the porosity of the sand. Show your work and be sure to show units of measure.

7. Clean the glassware by swishing it sideways in a pail of water to get rid of all the sand grains. Please keep sand out of the sinks and drains.

Wrap-Up ▏▏▏▶

1. Define porosity. _____

2. Why are sandy soils relatively porous?

_____

3. If your source of municipal or household water is groundwater, why is porosity of soil and rock very important to you?

_____

_____

4. What is the porosity of 200 mL of soil that can hold 80 mL of water? (Show your work below starting with the algebraic equation.)

5. A student placed a dry 150-mL sample of sandstone in a 500-mL beaker filled with 400 mL of water overnight. When the sandstone was taken out, the water level dropped to 350 mL. What is the measured porosity of the sandstone? (Be sure to show your work.)

🌐 CHAPTER 13—LAB 3: GROUNDWATER MOVEMENT

Introduction

Groundwater is often used as a safe and reliable source of water for domestic and agricultural uses. The ability of soil and underlying bedrock to absorb and convey groundwater depends on the physical properties of the material. Before beginning the procedure, you need to review the following important terms.

1. What do we call the percentage of open space within a soil?

_____

2. What is the ability of water to flow through a soil called?

_____

3. What term means the ability of a soil to hold back groundwater?

_____

4. What name is applied to the movement of groundwater upward into tiny spaces?

_____

Although you will be working with samples of plastic beads or stones that have been sorted into uniform sizes, these still provide a useful model of natural soils.

Objective

Investigations of the materials listed below will help you understand how the properties of capillarity, porosity, permeability, and retention relate to the size of particles in soil.

Materials

Plastic column (with an outlet and a wire screen in the bottom) supported by a ring stand and two clamps, 2 plastic 400-mL beakers, funnel, pinch clamp, graduated cylinder (100 mL or 500 mL), stopwatch, drying pan, 3 sizes of dry, sorted beads or rounded stones

Procedure

1. The plastic column must have a wire screen in the bottom to prevent particles from blocking

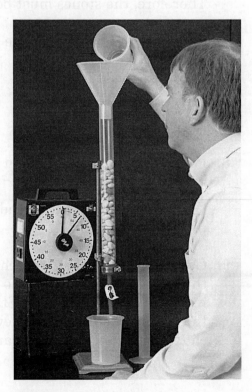

FIGURE 13-3.

the outlet. Also be sure the pinch clamp on the outlet is tight. Pour in a little water to see if it leaks. If there is no leakage, pour out the water and continue the lab.

2. Using a graduated beaker, measure 300 mL of the smallest dry particles. (If the beaker is glass, take care not to break it.)

3. Select one bead or a stone that is average in size and shape. Measure its dimensions.

    The greatest length of this stone is _____ mm.

    The thinnest diameter of the stone is _____ mm.

    Therefore, the average diameter is _____ mm.

    (Enter this measurement in Table 13-1.)

4. Lift the whole ring stand and column and then stand it upright on the floor. Hold a plastic funnel over the top of the column. Using the funnel, pour the stones into the column. *Pour slowly so you do not loosen the end cap.*

5. Close the pinch clamp watertight. Then, measure exactly 300 mL of water with a graduated cylinder.

6. Slowly pour the water into the column until it just rises to the top of the stones. Save the rest of the water for Step 8.

    How much water was left in the beaker? _____ mL

    Therefore, the stones must be holding how much water? _____ mL

7. Calculate the porosity of the stones. (Remember that the total volume is 300 mL.)

8. Pour the rest of the water in the beaker into the plastic column. Place an empty beaker under the column.

    (Step 9 is just for you to read. But be sure you understand it before you do the procedure in Step 10.)

9. Get ready to use a stopwatch to measure the time it takes for the water to drain through the particles, but do not do it yet. As the water drains, you will watch the water level in the column. Start the stopwatch when you release the clamp. Stop the watch when the water level reaches the bottom of the tube. You must watch the changing water level very carefully.

10. Carefully, but quickly, open the pinch clamp. How long does it take the water to drain through the particles to the bottom of the column?

_____

11. Use the graduated cylinder to measure the amount of water that came out. In addition, determine how much water was held back by the stones.

_____ mL came out, _____ mL left behind

TABLE 13-1. Student Data

Average Particle Size (mm)	Porosity (%)	Permeability Time (seconds)	Water Retention (mL)
Small			
Medium			
Large			

12. Empty the wet beads or stones into a drying pan. Be sure that all the particles fall out of the tube. (You may need to tap it.) Then carefully repeat this procedure with the medium and large stones.

13. Your teacher has set up small columns of sand to illustrate capillary action. Ask where you may observe these tubes. How does the size of the particles affect how high the water moves by capillarity?

_____

From your earlier learning, you should already have a good idea how the particle size affects each of these four factors: porosity, permeability, retention, and capillarity.

If any of your data seems to contradict what you have learned earlier, you have two choices. You can do that part of the experiment again. Or attach a sheet of notebook paper to your lab report with an explanation of how the data should look, and what may have caused your experimental results to be in error.

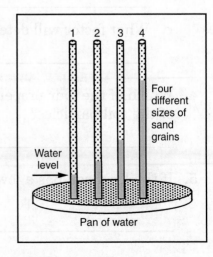

FIGURE 13-4. Observations of capillarity.

Wrap-Up

1. Which of the lines on this graph shows how the size of the particles affects the following properties?

   a. Porosity   _____

   b. Permeability rate   _____

   c. Permeability time   _____

   d. Water retention   _____

   e. Capillary movement   _____

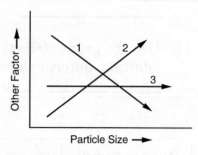

FIGURE 13-5.

2. What three soil properties of sediments actually do affect soil porosity?

   a. _____

   b. _____

   c. _____

3. Congratulations! You have purchased a farm in a rural area without public utilities. That means you will need to dig or drill a well to access groundwater. The properties of the soil will affect your ability to draw water.

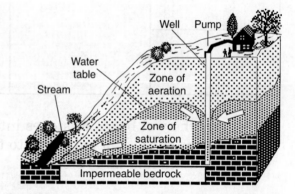

FIGURE 13-6. Profile view of an aquifer.

   a. What soil property will determine how quickly water will seep into your well from below the water table?

   _____

   b. What factor will affect how water moves up through the soil into the root zone?

   _____

   c. What factor will determine the total amount of water the soil holds?

   _____

4. Which of the four properties you observed in this lab does not depend upon the size of the soil particles?

   _____

5. If the local soil has a low permeability, what might you see that would indicate this condition?

   _____

   _____

Chapter 14
Glaciers

☀ CHAPTER 14—SKILL SHEET 1: LANDFORMS OF GLACIATION

There are alpine (mountain) glaciers and continental glaciers. There are still a few alpine glaciers in the northern Rocky Mountains. At least four great ice sheets, continental glaciers, covered New York during the past 2 million years. This activity will be done in several parts. The object is to help you become more familiar with a variety of landforms and other geologic evidence of continental glaciation. Perhaps this will enable you to identify similar features in your own surroundings.

Part I. Definitions of Glacial Landforms

Drumlin: A long, narrow hill composed of glacial till, which was shaped by moving ice. In New York, drumlins are steep on the north end and trail off to the south.

Erratic: A large rock or boulder moved by a glacier. Some erratics differ in rock type from the bedrock below them.

Esker: A sinuous (winding) ridge deposited by a stream in a tunnel under a glacier.

Kame: A small hill of sediment deposited by a stream, often at the edge of a glacier. This process is similar to the formation of a delta where a stream enters calm water.

Kettle: As shown in Figure 14-1, a closed depression in till, sometimes created as an ice block melted, leaving a hole. Some kettles fill with water to form kettle lakes.

Moraine: An elongated and irregular pile of unsorted till deposited by moving ice, usually at the side or at the end of a glacier.

Outwash: Sediments carried away from a glacier by meltwater. Outwash is usually stratified (layered) and sorted.

FIGURE 14-1. Kettle Lake on Cape Cod, Massachusetts.

FIGURE 14-2. Whaleback (roche moutonnée) on a golf course in Westchester County, New York.

Till: An immature soil, with little organic material, composed of unsorted sediment that has been moved into place by a glacier.

Polish: A bedrock surface smoothed by glacial abrasion. If fresh, it may reflect light.

Striations: Scratches on bedrock formed when a glacier dragged embedded rocks over a bedrock surface.

Grooves: Long channels or furrows carved by rocks dragged under a moving glacier.

Whaleback: A rock knob shaped and plucked by moving ice. The sharper, plucked end is usually on the end to which the glacier advanced. (These are also known by their French name roche moutonnée.)

Part II. Identify Glacial Features

For this part, use Figure 14-3 and the definitions in Part I. Write in the correct name of the eight features shown on the diagram below, A through G. (The last four features defined above are too small to show on this diagram.)

A. _____   E. _____

B. _____   F. _____

C. _____   G. _____

D. _____   H. _____

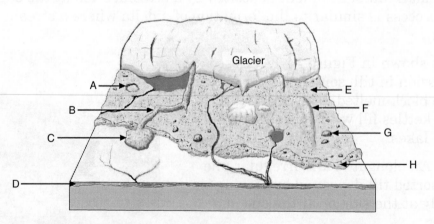

FIGURE 14-3. A glacier recedes when the rate moving back due to melting is greater than the forward movement of the ice.

Part III.

Identify features A–H above on the images in Figure 14-4. Each photograph illustrates one feature. On the line near its photograph, label each landform with its name. Use your answers from Part 1. One example has been done for you to show you how to label these images.

FIGURE 14-4.

Part IV. Shape and Symmetry

Landforms shaped by glaciation often show a characteristic size, shape, and symmetry. The sketches in Figure 14-5 represent glacial features. An object with bilateral symmetry can be split down the middle and is the same on each side, such as a spoon or the letter O. An object with radial symmetry is similar in all directions, like a wheel. From the information provided in Part I, identify each landform labeled in Figure 14-5.

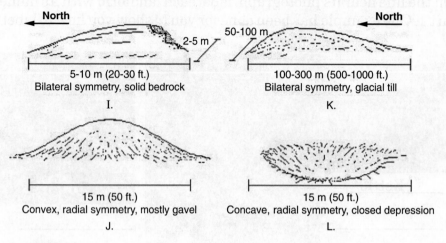

FIGURE 14-5.

I. _____   **K.** _____

J. _____   **L.** _____

M. Whalebacks are similar in shape to drumlins. How do they differ in shape?

_____

Part V. Identification of Glacial Features

The following items describe various glacial landforms. Identify each by name.

N. Parallel lines on bedrock. In New York they are often oriented north-south.

_____

O. A soil of mixed particle sizes from clay to boulders with little organic content.

_____

P. Although some of these objects weight many tons, they may have been pushed, dragged, and carried many miles by the moving ice.

_____

Q. This kind of bedrock surface is worn so smooth it reflects light. _____

R. By what two characteristics do sediments transported by meltwater differ from sediments deposited directly by glacial ice?

_____

_____

CHAPTER 14—SKILL SHEET 2: GLACIAL FEATURES OF NEW YORK STATE

The landscapes of New York State have unusual features that puzzled geologists for many years. Perhaps you have seen them, too. You can see huge boulders perched on high ridges. These boulders are sometimes composed of rock types unlike the bedrock on which they are found. New York has many immature and unsorted soils composed of transported material. Geologists wondered about oddly shaped mounds of sand and gravel throughout New York State as well as rounded, grooved, and polished bedrock surfaces.

Harvard geologist Louis Agassiz proposed an explanation of the origin of these features. He had observed similar features near current glaciers in the Alps of Switzerland. Agassiz suggested that these features were created when a great continental ice sheet covered most of North America.

This idea met with skepticism at first. The idea of ice covering half of North America seemed unbelievable. However, mounting evidence showed that there could be no other explanation.

In fact, geologists have found evidence of several major glacial advances over the past 2 million years. Each advance started when more snow fell over many winters than melted in the summers. (See Figure 14-6.)

FIGURE 14-6. The limits of continental glaciation in North America.

As the snow built up, it compressed underlying layers of snow changing them into ice. This pressure also caused the ice to flow outward from the center of accumulation. Although ice is generally considered a solid, we see it bend and flow under long-term stress. Under the force of gravity, continental glaciers spread southward from eastern Canada and covered New York State. (See Figure 14-7.)

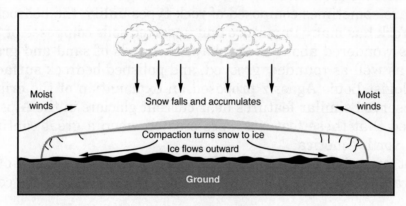

FIGURE 14-7. Profile of a continental glacier: Ice flows outward and downhill in response to gravity.

The most recent glacial advance, about 20,000 years ago, covered nearly all of New York State. When the climate turned warmer, the glacier gradually melted back to the north, leaving even the St. Lawrence valley free of ice by about 10,000 years ago. (See Figure 14-8.)

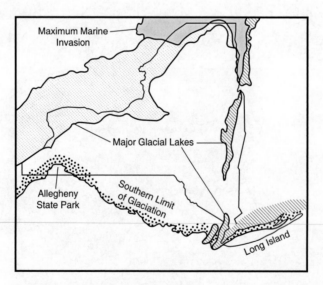

FIGURE 14-8. The limits of glaciation in New York State.

1. According Figure 14-8, what part of New York State was *not* covered by the most recent glaciers?

2. Why did the ice flow southward from Canada?

_____

3. Where did Agassiz observe glacial features before he came to America?

_____

Glacial meltwater deposits have become the major source of New York's most economically important geological resource: sand and gravel, which are used for building roads and landscaping.

Soil profiles in New York are also unusual. A soil that formed in place is called a residual soil as shown in Figure 14-9. They have horizons (layers) of weathered rock over broken up bedrock. However, most New York soils contain a mix of weathered sediments, but the layer of deeply weathered and broken bedrock below is usually missing. The soils of New York are composed of sediments transported by the glaciers. Many of these sediments have been stripped from one location by the moving ice and later deposited over solid rock. Some of this till has been pushed and dragged many tens of miles to the south

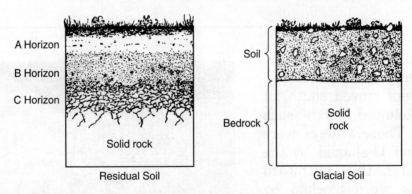

FIGURE 14-9. Many New York soils are transported soils.

4. How do materials deposited directly by the ice differ from sediments deposited by meltwater?

_____

_____

5. Why is the lowest soil horizon (weathered bedrock), absent from many New York soils?

_____

_____

A moraine is formed as a glacier pushes mounds of till (glacial sediment) like a bulldozer. If the glacier is melting back at the same rate that the ice flows forward, the ice front stays in the same place. However, the continuing conveyor-belt-like motion of the ice pushes more and more sediment into a pile at the end of the glacier. A **terminal moraine** is a ridge that marks where a glacier stopped advancing. Long Island was created in this way by a glacier. Two terminal moraines form the "backbone" of Long Island, giving it a

Y shape, as shown in Figure 14-10. The northern shore of Long Island is built of mounds of glacial till that contain particles from clay size up to boulders, mixed together. Therefore, many north shore beaches are very rocky with steep bluffs. The south shore is very different. It is made of the fine sand washed out of the glacier by meltwater. Most south shore beaches are more flat and sandy.

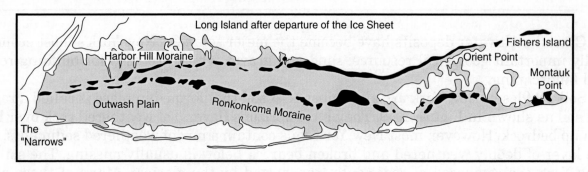

FIGURE 14-10. Moraines of Long Island.

6. Why are bedrock exposures absent on most of Long Island?

Glaciers also created the Finger Lakes of western New York State. When the glaciers moved south, at first the ice followed north-south stream valleys. These valleys were made deeper and U-shaped by the moving ice. Later, the southward flowing streams were blocked by till deposits (recessional terminal moraines). Some of the dammed north-south valleys filled with water, forming the Finger Lakes. (See Figure 14-11.) Other valleys are dry, clearly showing the characteristic U-shape of glaciated valleys.

No matter where you live in New York State, it is likely that you can observe glacial landforms in your

FIGURE 14-11. Carved deeper by glaciers, the Finger Lakes show the direction of ice movement.

area. Perhaps you have seen erratic bounders left high above a stream or river, such as Shelter Rock on Long Island. Exposed bedrock surfaces may be rounded, polished, and scratched. Gravel pits and road cuts reveal unsorted glacial till and layered meltwater sediments. Even the fabric of the land (large-scale patterns) such as the alignment of ridges and valleys is probably a result of glacial erosion and deposition.

7. On another piece of paper, make a profile sketch of a steep valley that was eroded by a stream and a second profile sketch of a similar valley cut by a glacier. Label each profile.

☀ CHAPTER 14—LAB 1: INQUIRY INTO GLACIAL MOVEMENT

You can model glacial flow with Ooblek. Your teacher may provide Ooblek, or you can make Ooblek by mixing $1\frac{2}{3}$ cups of water with pound of cornstarch. Adding food coloring is optional.

For this lab, you will devise an experiment with Ooblek that you can use to measure the relationship between flow rate of the glacier and slope. After you have determined your procedure, have your teacher check it.

After you have conducted your experiment, write a report using the form given in Chapter 10 or one suggested by your teacher.

My Notes

Chapter 15
Landscapes

 CHAPTER 15—SKILL SHEET 1: LANDSCAPES OF NEW YORK STATE

Landscapes can usually be classified as plains, plateaus, or mountains. Of these, plains have the least relief and mountains have the most relief. New York State's landscape is remarkably diverse. See Figure 15-1. As you travel from one landscape region to another, landscape boundaries are often remarkably clear. For example, the Catskill escarpment (the eastern boundary of the Catskill Mountains) is visible to the west of the New York State Thruway over much of the distance from New York City to Albany.

Landscapes regions develop due to differences in geology and climate. The climate in New York State is relatively uniform: continental and humid. However, New York does have a great variety of rock types and geological structures. The following text describes nine landscape regions of New York State.

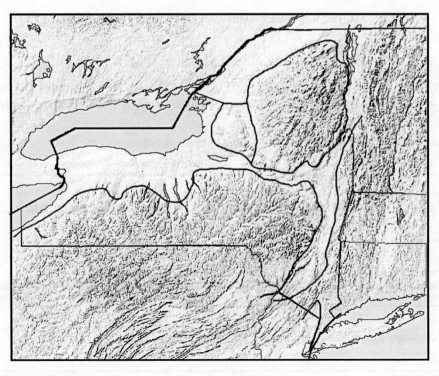

FIGURE 15-1. Shaded relief map of New York State.

Long Island is a part of the **Atlantic Coastal Plain**. It was formed when the southward-moving continental glaciers pushed two long ridges of glacial till into place. See Figure 15-2. The southern half of the island is composed of fine sand transported by meltwater from the glaciers. Except at its extreme western end, there are no bedrock exposures on Long Island. The south shore of Long Island is protected by a series of barrier islands. These islands are composed of sand washed along the shore by waves and ocean currents.

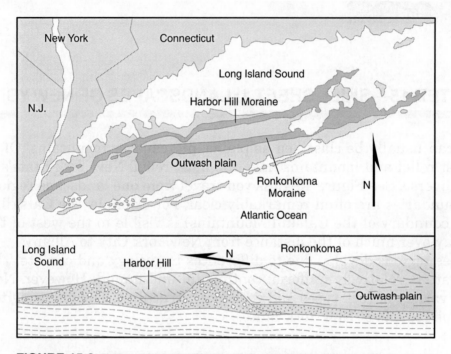

FIGURE 15-2. Two terminal moraines form the backbone of Long Island.

To the west, the south shore barrier islands end at the sand spits that guard the entrance to New York Harbor. This large, deep, and well-protected harbor helped to make New York City the most important center of trade and finance in the world.

1. What two factors usually influence landscape development?

_____

2. Which of these two factors is most variable within New York State?

_____

3. What agent of erosion and deposition formed Long Island?

_____

4. What processes are now modifying the shoreline of Long Island?

_____

The **Newark Lowlands** is a region of red sedimentary rock, mostly shale and sandstone, located west of the lower Hudson River. It is bounded on the east by the Palisades, a thick intrusion of basaltic igneous rock that forms the great cliffs along the Hudson River from New York City north to Haverstraw. See Figure 15-3. The Palisades intrusion of magma occurred about 200 million years ago during the Triassic Period when dinosaurs first appeared on Earth. At that time, stretching occurred as the Atlantic Ocean was just beginning to open.

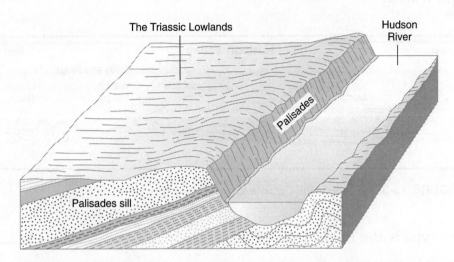

FIGURE 15-3. The Newark Lowlands are bounded on the east by the Palisades intrusion.

The **Hudson Highlands and Taconic Mountains** is a region of very old bedrock and complex geologic structures. See Figure 15-4. Many of the rocks are high-grade metamorphic rocks exposed by extensive erosion of an ancient mountain range. These mountains were probably pushed up by the collision of the ancient continents of North American and Africa about 400 million years ago.

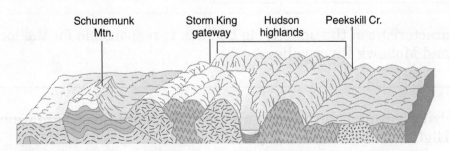

FIGURE 15-4. The Hudson Highlands are composed of ancient igneous and metamorphic rocks.

The Hudson and Mohawk Rivers cut the **Hudson and Mohawk Lowlands** mostly through relatively weak sedimentary rocks. These valleys provided important low-level trading routes to the American interior in colonial times. These corridors were later used for the Erie Canal and the New York State Thruway.

West of the Hudson Valley is the **Catskill Mountains and Allegheny Plateau**. Thick layers of flat lying sedimentary rocks underlie this region. The layers were deposited as a delta in an inland sea about 400 million years ago. The sediments probably originated in the high mountain range that existed where we now find the Hudson and Taconic Highlands. After a regional uplift, the Catskill delta was deeply eroded by streams and rivers. The topographic relief of this landscape region decreases westward as the Catskill Mountains progress into the Appalachian Plateau of southwestern New York State. See Figure 15-5.

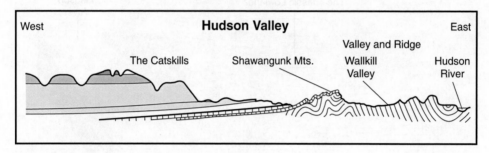

FIGURE 15-5. Profile from the Catskills to the Hudson River.

5. What rock type is the Palisades Sill? _____

6. Folded and faulted crystalline (igneous and metamorphic) rocks underlie most mountains. In what two ways are the Catskills different from most mountains?

_____

_____

7. The rocks of the Catskill Mountains were originally deposited as what kind of landform?

_____

8. What characteristic of the underlying bedrock is responsible for the location of the Hudson and Mohawk River valleys?

_____

_____

9. What global event probably pushed up the high mountains where we now find the Hudson Highlands and Taconic Highlands?

_____

_____

10. Where might you look in New York State if you wanted to find fossils of dinosaurs?

_____

The **Erie-Ontario Lowlands** borders the Great Lakes. Layers of sedimentary rocks in this region dip gently to the south. The glaciers left many characteristic features that dominate the landscape including moraines, drumlins, eskers, and kames. The glaciers

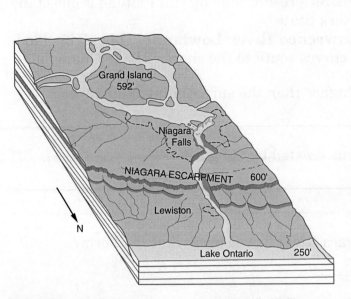

FIGURE 15-6. Niagara Falls formed where the outlet of the upper Great Lakes crossed the Niagara Escarpment.

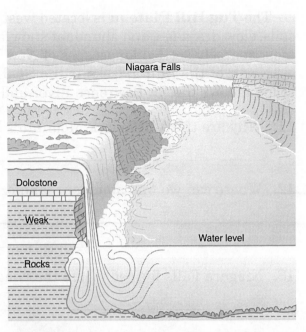

FIGURE 15-7. Niagara Falls is the result of the Niagara River running over a resistant cap rock.

also deepened and dammed the valleys now occupied by the Finger Lakes. Niagara Falls formed where the Niagara River runs over a layer of resistant dolostone. See Figure 15-6.

The Lockport Dolostone forms the cap rock of Niagara Falls as well as a bluff, or ridge, known as the Niagara Escarpment. (See Figure 15-7.) The falls began where the river reached the crest of the escarpment about 15,000 years ago. Erosion is causing Niagara Falls to move steadily upstream at the remarkable rate of about 2 meters per year.

The **Adirondack Mountains** are the only true mountain landscape of New York State. They are an extension of the Grenville Province, a part of the ancient Precambrian core of North America known as the Canadian Shield. The Adirondacks are a great dome, pushed up in the middle to expose a core of ancient plutonic and metamorphic rocks. New York's highest point, Mount Marcy, at 1629 meters (5344 feet) is located in the Eastern Adirondacks.

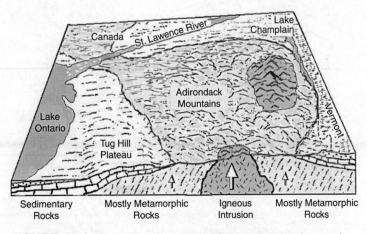

FIGURE 15-8. The Adirondack Mountains were pushed up and overlying sedimentary rocks have been eroded away.

The **Tug Hill Plateau** is located west of the Adirondacks. This is a hilly region of poor soils, slow drainage, and abundant snowfall. As a result, the Tug Hill Plateau is one of the most sparsely settled landscapes of New York State.

North of the Adirondacks is the **St. Lawrence River Lowlands**. This plains landscape follows the St. Lawrence River and curves south to the shore of Lake Champlain.

11. Why are the Adirondack Mountains higher than the surrounding landscapes?

_____

12. Where can we find abundant landforms created by the glaciers?

_____

_____

13. Niagara Falls formed where the Niagara River flowed over what landform?

_____

14. What specific type of rock holds up Niagara Falls? _____

15. Why do so few people inhabit the Tug Hill Plateau?

_____

16. What page of the *Earth Science Reference Tables* has a landscapes map? _____

17. Why do Figure 15-9 and the *Earth Science Reference Tables* have slightly different landscape boundaries?

_____

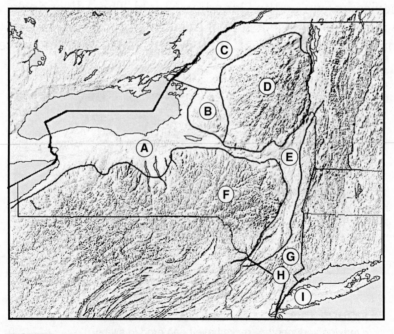

FIGURE 15-9. Landscapes regions of New York State.

18. Use the *Earth Science Reference Tables* to label the nine landscape regions labeled in Figure 15-9.

A. _____

B. _____

C. _____

D. _____

E. _____

F. _____

G. _____

H. _____

I. _____

19. Region G, the Hudson and Taconic Highlands, is difficult to classify because of its complexity. This region shows characteristics of mountain, plateau, as well as plains landscapes. What letter labels the only true mountain landscape in New York State? _____

20. New York's plateau landscapes are lettered _____

21. The plains landscapes are lettered _____

22. According to the *Earth Science Reference Tables*, New York's Catskills and Allegheny Plateau are a part of what much larger landscape region?

_____

23. The Adirondacks are shown in the *Earth Science Reference Tables'* map as an extension of what major landscape?

_____

24. Which of these four states do you think has the fewest landscape regions?

   a. California   b. Colorado   c. Delaware   d. Virginia

   _____

Another characteristic of landscape regions is their similar **stream patterns**. See Figure 15-10. For example, where the bedrock is uniform or the layers are flat, a branching pattern called **dendritic** drainage develops. Volcanoes or uplifted dome mountains often display **radial** drainage. This looks something like the radiating spokes on a wheel. **Rectangular** drainage, where many streams meet at right angles, often occurs where there are many faults or joints (stress cracks) in bedrock. An **annular** pattern suggests concentric circles where streams flow around a topographic dome but are held in by upturned layers before they exit outward.

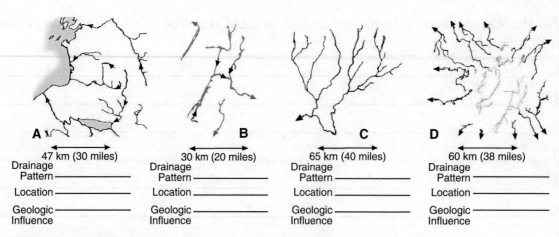

FIGURE 15-10. Four stream patterns within New York State.

25. The four stream patterns in Figure 15-10 can be found on the map in Chapter 12—Skill Sheet 2: Watersheds. Use that map for this problem. For each example, A, B, C, and D in Figure 15-10, state the type of drainage pattern, identify the location of these streams on the Chapter 12 map, and tell what characteristic of the local geology has caused this pattern to appear.

A. _____

B. _____

C. _____

D. _____

26. What word describes the general shape and characteristic features of a land area?

_____

27. In which New York State landscape region do you live?

_____

Chapter **16**
Oceans and Coastal Processes

 CHAPTER 16—LAB 1: WHAT IS ON THE BEACH?

Introduction

Sediments collected from different locations have unique characteristics. For example, the name "sand" tells us the general size of the particles. However, sands vary greatly in their mineral composition, surface texture, sorting by size, and other characteristics.

Objective

Use a hand lens and other materials as provided to record your observations of at least eight sediment samples. See Figure 16-1. Pick any eight of the samples available and neatly record your observations in Table 16-1. (Your teacher will help you understand the column headings.)

FIGURE 16-1. Examining sediment with a hand lens.

Materials

Samples of different kinds of sand, hand lens, metric ruler, magnet in a plastic bag

Procedure

TABLE 16-1. Beach Sand Data Table

Sample Letter	Overall Color	Colors of the Grains*	Degree of Sorting*	Relative Angularity*	Other Characteristics	Geographic Source of Sediment

*Use a magnifying lens.

1. Measure the size of a typical grain of sand. Pick one sediment sample that is well sorted and has grains of a typical sand size. Place one average grain next to the millimeter scale on the *Earth Science Reference Tables* or a metric ruler. Look at it with a hand lens and estimate the average size (diameter) of this grain of sand.

2. What minerals can you identify in these materials?

3. In what two (2) ways does a sand sample differ from most soils?

4. According to the *Earth Science Reference Tables,* what is the size range of sand particles?

5. What two agents of erosion and deposition commonly deposit sediments sorted by size?

6. What agents of erosion do not sort grains by size?

CHAPTER 16—SKILL SHEET 1: GRAPHING THE TIDES

The water level in the oceans changes each day in a cycle. This change in sea level is called the tides. The tides are caused by the gravitational attraction of the moon and, to a lesser extent, the sun.

FIGURE 16-2. The tides are cyclic changes in sea level.

Ocean tides often extend into streams that flow into the ocean. The tides affect the level of the Hudson River as far north as Albany. The data in Table 16-2 shows the changing water level of the Hudson River at Peekskill, NY, over a period of 42 hours. Use this data to construct a graph showing the changing water level. The data points must be located both above and below a "0" mean sea level line that runs across the middle of your graph.

TABLE 16-2. Changing Water Level of the Hudson River at Peekskill, New York

Date	October 20 A.M.											
Time	Midnight	1	2	3	4	5	6	7	8	9	10	11
Height (m)	0.8	0.7	0.5	0.2	−0.2	−0.6	−0.7	−0.65	−0.4	0	0.3	0.5
Date	October 20 P.M.											
Time	Noon	1	2	3	4	5	6	7	8	9	10	11
Height (m)	0.6	0.6	0.5	0.2	−0.1	−0.4	−0.5	−0.5	−0.4	−0.1	0.1	0.3
Date	October 21, A.M.											
Time	Midnight	1	2	3	4	5	6	7	8	9	10	11
Height (m)	0.45	0.5	0.45	0.3	0	−0.3	−0.45	−0.5	−0.45	−0.3	−0.1	0.1
Date	October 21, P.M.											
Time	Noon	1	2	3	4	5	6					
Height (m)	0.3	0.4	0.4	0.3	0.1	−0.1	−0.3					

Wrap-Up

1. How can you tell that this change is cyclic?

2. How many complete cycles are shown on your graph? _____

3. What is the time of the first afternoon high tide to the nearest 15 minutes?

   _____

4. What is the tidal range in the first tide cycle? (The range is the difference between the highest and the lowest values.)

   _____

5. Is the tidal range increasing or decreasing? _____

6. What is the *average* time from one high tide to the next? _____

7. What causes the tides?

8. Why might it be important for a person to know the time and height of the tides?

9. The sun is much larger than the moon by about 1000 times. So why does the moon have more affect on the tides? (There are at least two reasons.)

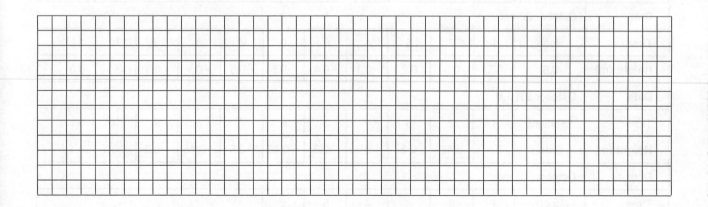

Chapter 17
Unraveling Geologic Time

 CHAPTER 17—LAB 1: RELATIVE AND NUMERICAL AGE

Introduction

There are two ways to express the age of things. If you specify an age by comparison with other objects (younger or older), you are expressing a relative age. A numerical, or a quantitative, age is expressed in specific units of time, such as seconds or years. Numerical measures are usually more exact than relative measures, although this is not always the case. Quick as a flash describes quite a precise time period. Numerical values always include numbers and units of measure, such as 5 hours and 20 seconds, or about 1 million km.

FIGURE 17-1. Most objects have clues to help you infer an age.

Numerical age is sometimes called absolute age. However, this does not mean the age is known with absolute certainty. No measurement is perfect. Using a measuring instrument such as a clock or a stopwatch can reduce errors. However, no matter how carefully an observation or measurement is made, better instruments or better techniques can always produce greater accuracy. A small margin of error, therefore, means that you are confident that your observation, measurement, or estimate is very close to the actual age of the object.

Objective

To estimate the numerical age of different objects, establish a margin of error and rank the objects from oldest to youngest.

Procedure

Fill in Table 17-1 using the 10 objects that your teacher will provide. Use the margin of error to express a time period that you are reasonably confident includes the actual age of the object you are observing. For example, you might estimate the age of a student in the first grade to be 6 years old ± 1 year just knowing the educational level of the student.

TABLE 17-1. Numerical Age of Objects

Name of Item	Numerical Age	Estimated Margin of Error	Method Used to Date the Object
1		±	
2		±	
3		±	
4		±	
5		±	
6		±	
7		±	
8		±	
9		±	
10		±	

In Table 17-2, list the 10 objects in Table 17-1 in order, from the youngest at the top, to the oldest at the bottom. In this procedure, you are listing these objects by relative age.

TABLE 17-2. Relative Age of Objects

Youngest

Oldest

Wrap-Up ▶

1. Unlike a relative value, a numerical value always includes what two things?

2. Which of the 10 objects you worked with was the easiest to assign an age?

3. Is the time interval between the age of objects in Table 17-2 uniform?

4. What is the purpose of the margin of error? _____

5. How does the margin of error indicate the degree of the uncertainty of your estimate?

6. Why is it impossible to have a 0 margin of error for any measurable value?

7. Label the following as relative or numerical age.

    a. Very young _____

    b. 5 minutes old _____

    c. About a week old _____

    d. Oldest of all _____

    e. Younger than I am _____

    f. Immeasurably old _____

    g. A fraction of a second old _____

    h. Old enough to vote in a national election _____

8. If someone wanted to measure the exact age of a house, how exactly could it be measured? What would limit the precision of this measurement?

_____

_____

_____

9. In the space below, construct a timeline to show the age of all of the objects except any rocks. Your timeline must have a scale. (Use a ruler or meterstick.) That is, the same distance everywhere along the scale must represent the same length of time. Place the oldest objects to the left, moving toward the youngest on the right.

10. (Optional) At the scale you used in question 9, how long would the time line need to be to show the age of Earth?

_____

_____

CHAPTER 17—SKILL SHEET 1: ESTABLISHING SEQUENCE

Geologists can interpret events of the geologic past by making observations of bedrock outcrops and structures. When you see evidence of so many changes, you might think that these changes were more active in the past then they are today. However, geologists have found that this is not the case. The **Law of Uniformitarianism** states that the processes that operated in the geologic past are similar to those we observe today. Stated another way, "The present is the key to the past."

Observations of rock outcroppings can lead to an inferred sequence of events. Although this technique does not yield a numerical (quantitative) age, it does establish the order of events from youngest to oldest: that is, relative age.

The three sections in Figure 17-2 show how younger sediments are always deposited on top of older sediments. These diagrams illustrate the **Law of Superposition**, which states that in any outcrop, the oldest layers of rock are generally those on the bottom. **Original horizontality** means that layers of sediment are deposited in level layers. If we see them tilted or distorted, this has occurred after deposition.

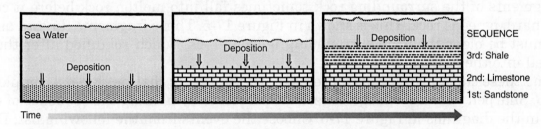

FIGURE 17-2. In most cases, the first layers deposited end up on the bottom.

In some outcroppings, folding or faulting has caused overturning, resulting in younger layers being beneath older layers. In these unusual instances, you may be able to find younger fossils that are under older fossils, or you may find graded bedding that shows a gradual change from larger particles on top, toward smaller particles on the bottom. This would show that the rocks had been turned upside down. However, if you see no evidence of overturning, assume that the oldest layers are on the bottom.

Volcanoes and meteor impacts can create time horizons. Violent volcanic eruptions can cover vast areas with a thin layer of volcanic ash that settles very quickly in terms of geologic time. When meteors collide with Earth, they can quickly spread trace minerals at the surface over vast areas.

Figure 17-3 shows that rocks must be in place before they are cracked, tilted, folded, or faulted. It is only logical that the rocks must form before they are changed by these events.

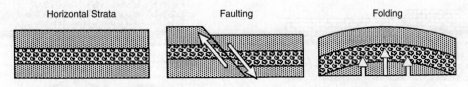

FIGURE 17-3. Faults and folds are younger than the rocks in which we find them.

Intrusions of molten rock occur after the formation of the surrounding rock. A zone of baked rock, called contact metamorphism (shown here by the short, black lines), forms next

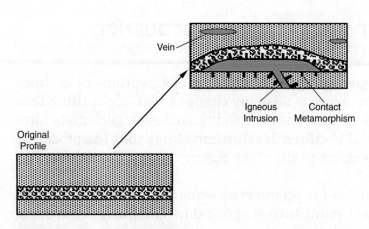

Vein

Original
Profile

FIGURE 17-4. Igneous intrusions are always younger than the rocks they were injected into.

Igneous
Intrusion

Contact
Metamorphism

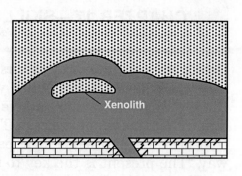

Xenolith

FIGURE 17-5. A xenolith is a solid rock fragment that has fallen into molten magma.

to the new igneous rock at the time of intrusion. **Intrusions** are completely within the ground (internal). A **vein** is a small intrusion. Veins are also younger than the rock in which they are found. (See Figure 17-4.) **Extrusions** come out at the surface. They are external.

Fragments of the surrounding rock sometimes fall into molten rock before it cools to form a hard, igneous rock. This is shown in Figure 17-5. These fragments (known as xenoliths) must be older than the surrounding igneous rock, which solidified after the fragment fell into the magma.

Using these clues, a geologist can reconstruct the events that produced a complex rock outcrop. Sharpen your knowledge of geologic processes by naming the sequence of events shown in the diagrams in Figure 17-6. Choose the events from the following list: Deposition, Intrusion, Extrusion, Tilting, Folding, Faulting, and Erosion.

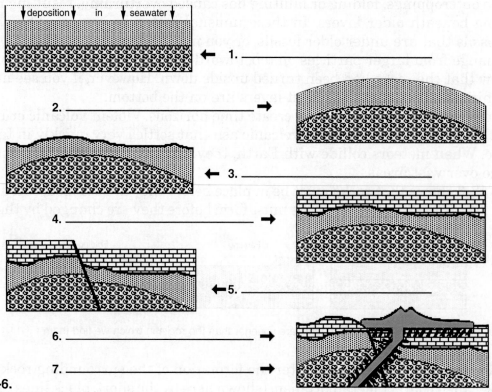

deposition   in   seawater

1. _____

2. _____

3. _____

4. _____

5. _____

6. _____

7. _____

FIGURE 17-6.

1. _____

2. _____

3. _____

4. _____

5. _____

6. _____

7. _____

Make a list of the sequence of geological events necessary to produce the profile in Figure 17-7.

8. _____

9. _____

10. _____

11. _____

12. _____

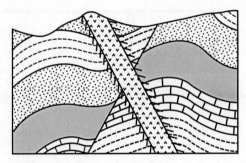

FIGURE 17-7.

 Wrap-Up

13. Define horizontal. _____

14. What is the term for a break in rocks along which movement has occurred?

_____

15. What is the geologic term for a bend in rock layers? _____

16. Explain uniformitarianism? _____

17. What is a vein? _____

18. How does intrusion differ from extrusion?

_____

19. What do geologists mean by the Law of Superposition?

_____

20. What two events can spread geologic time markers over large areas of Earth?

_____

21. Which of these events is most likely to occur in just a few seconds? (_Hint:_ Use your _Earth Science Reference Tables._)

_____

22. An intrusion of granite is found to contain large inclusions of hornfels. The surrounding bedrock is shale and sandstone. Where did the hornfels come from?

_____

_____

_____

CHAPTER 17—LAB 2: FOSSIL CORRELATION

Introduction

Correlation is matching. Rocks can be correlated by rock type or by age. In this laboratory you will be correlating the age of widely separated rock layers by using **index fossils**. The best index fossils are the remains of organisms that lived over a large geographic area, but were alive for a relatively brief time in geologic history. See Figure 17-8.

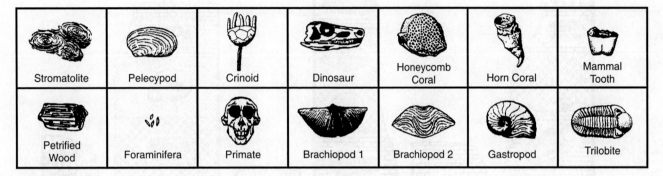

FIGURE 17-8. Index fossils (listed by their modern names).

Objective

To establish a relative scale of geologic time.

Procedure: Part I

Use the six widely spaced outcrops in Figure 17-9 to establish a relative scale of geologic time. In Table 17-3 on page 175, list the 14 index fossil organisms in their correct time sequence, from the oldest at the bottom to the youngest at the top. (In some cases, there will be two fossils of the same age; be sure to list both.) A dark, irregular interface between the layers shows a buried erosion surface (unconformity) where one or more layers are missing.

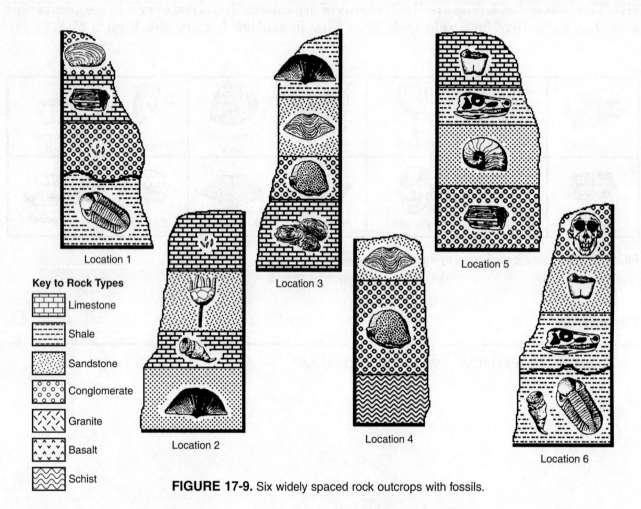

FIGURE 17-9. Six widely spaced rock outcrops with fossils.

Figure 17-10 provides the structural symbols used in the profiles in the rock outcrops used in this lab.

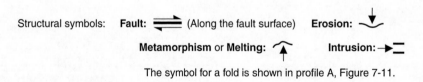

FIGURE 17-10. Structural symbols.

In Table 17-3 write in the appropriate fossil names with the oldest at the bottom to youngest at the top.

LABORATORY MANUAL

TABLE 17-3. Age of Index Fossils

Era	Period		Index Fossil(s)
CENOZOIC	Quaternary	12.	
	Tertiary	11.	
MESOZOIC	Cretaceous	10.	
	Jurassic	9.	
	Triassic	8.	
PALEOZOIC	Permian	7.	
	Carboniferous	6.	
	Devonian	5.	
	Silurian	4.	
	Ordovician	3.	
	Cambrian	2.	
PRECAMBRIAN	(Proterozoic)	1.	

1. Which organism is the same age as the pelecypod? _____

2. Does this procedure give a numerical or a relative age? _____

3. Where are the oldest fossils usually found in an outcrop?

   _____

4. What event seems to have destroyed a part of the fossil record in two of the outcrops?

   _____

5. What fossil(s) is/are missing in Location 1? _____

6. Of these fossil organisms, which organism is the most complex? _____

7. In any undisturbed outcrop, where are the most complex fossils found?

   _____

8. Why are there no fossils in the bottom layer at Location 4?

   _____

9. What are index fossils used for? _____

   _____

10. What properties make a fossil a good index fossil? _____

   _____

11. What kind of rocks are fossils usually found in? _____

Part II

The reference profile in Figure 17-11 shows seven fossils in their correct order within the Paleozoic Era. (The Paleozoic was the dawn of modern life-forms.) Sketch and name the structural change that could produce the features and sequence of fossils shown in each profile, A through I. The first one has been completed to show you how to indicate a fold. The other structures should also be shown according to the symbols in Figure 17-10.

FIGURE 17-11.

REFERENCE SEQUENCE

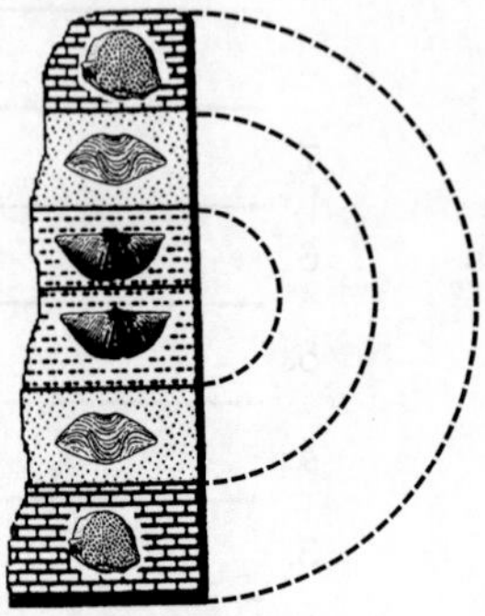

A. STRUCTURE: **Fold**

B. STRUCTURE: _____

C. STRUCTURE: _____

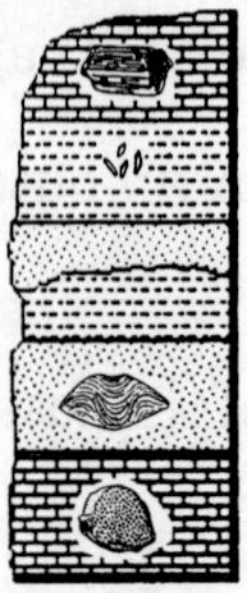

D. STRUCTURE: _____

E. STRUCTURE: _____

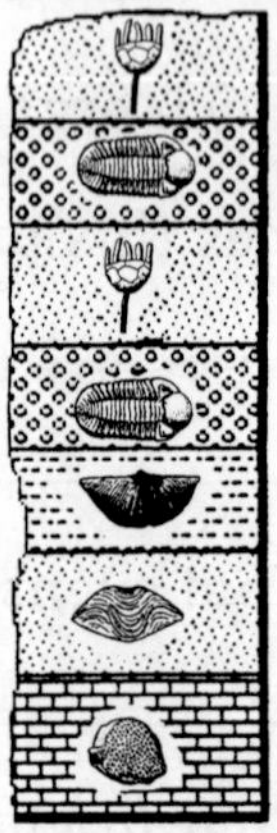

F. STRUCTURE: _____

G. STRUCTURE: _____

H. STRUCTURE: _____

I. STRUCTURE: _____

Key to Rock Types

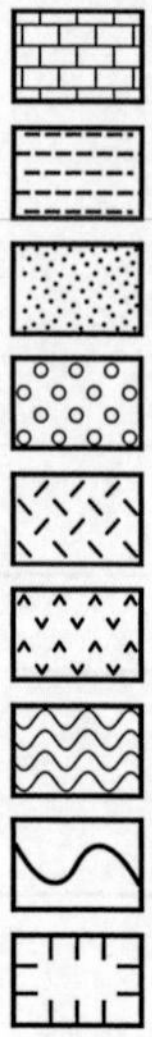

- Limestone
- Shale
- Sandstone
- Conglomerate
- Granite
- Basalt
- Schist
- Erosion Surface
- Contact Metamorphism

12. In what environment did most of these fossil organisms live?

_____

13. Humans could become or make excellent index fossils. What characteristics of humans might make them excellent index fossils?

_____

_____

14. How does the sequence of layers in a fold differ from the sequence in a fault? (*Hint:* Compare sections A and F.)

_____

_____

15. What events destroy part of the geologic record? (There can be three to four possible answers.)

_____

16. Which process always causes a profile to become thinner? _____

17. The distortion of rock layers by folding and faulting are often a result of large-scale global changes. What can cause these changes?

_____

18. What are two limitations of using fossils to interpret geologic structures?

_____

_____

19. What do scientists call the process by which organisms have changed through geologic history?

_____

20. How do scientists think life may have begun on the Earth?

_____

21. Some portions of the profiles shown on page 176 are upside down. Outline each upside-down part of these profiles with a dotted box. Be sure to mark all the upside-down portions of these profiles.

My Notes

CHAPTER 17—LAB 3: GEOLOGIC PROFILES

Introduction

Geology is sometimes called a backward science. In other fields of science, such as chemistry or physics, scientists perform experiments to find out the results of the experiment. However, in geology, scientists look at the results of millions of years of geological events (the "experiment," so to speak) and then attempt to determine what happened to produce the features that they are observing.

Objective

To interpret and re-create geologic profiles.

Procedure

Figure 17-12 is a key to the rock types used in Figures 17-13 and 17-14. Carefully examine diagrams A and B In Figure 17-13. For each diagram, the number next to the letter will tell you how many distinct geologic events were required to make the complete profile. (A process may be repeated; for example, deposition of several layers may be followed by erosion and then another period of deposition. That is three events.) Write the events needed next to each profile diagram.

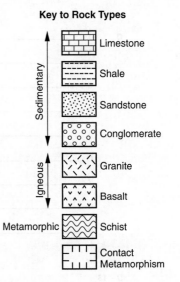

Key to Rock Types

FIGURE 17-12. Key to rock types.

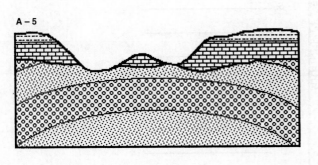

A – 5

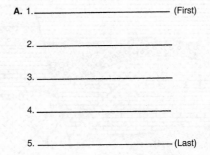

A. 1._____ (First)

2._____

3._____

4._____

5._____ (Last)

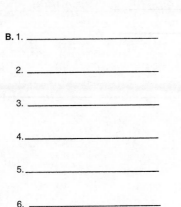

B. 1._____

2._____

3._____

4._____

5._____

6._____

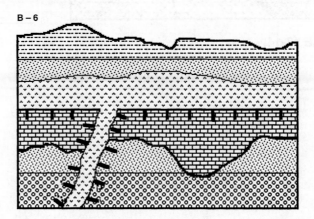

B – 6

FIGURE 17-13.

Select the events for each profile from the following list: Deposition, Erosion, Intrusion, Extrusion, Tilting, Folding, Faulting, Metamorphism, Melting, and Solidification (Crystallization)

For the profiles C and D in Figure 17-14, you should understand that a single feature may involve more than one event in a specific sequence. For example, igneous rocks form by melting followed by solidification. However, events that occur at the same time, such as an intrusion causing metamorphism, should be considered a single event. When you hand in your paper, you will be asked to choose the best sequence of events from four choices for each profile A–D.

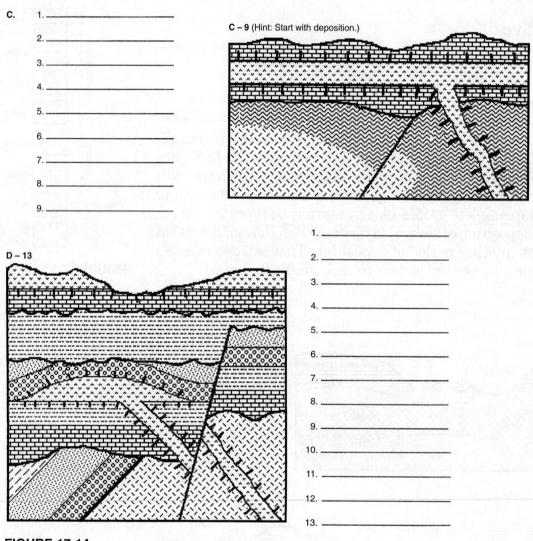

C.   1. _____

2. _____

3. _____

4. _____

5. _____

6. _____

7. _____

8. _____

9. _____

C – 9 (Hint: Start with deposition.)

1. _____

2. _____

3. _____

4. _____

5. _____

6. _____

7. _____

8. _____

9. _____

10. _____

11. _____

12. _____

13. _____

D – 13

FIGURE 17-14.

Draw neat geologic profiles to show the following three sequences of events. Please use the key in Figure 17-12.

E. 1. Deposition

   2. Tilting

   3. Erosion

F. 1. Deposition

   2. Faulting

   3. Erosion

   4. Deposition

G. 1. Solidification

   2. Erosion

   3. Deposition

   4. Extrusion

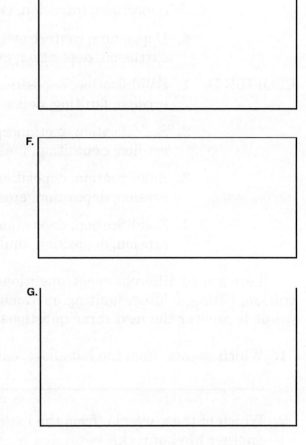

Write the event names (such as "deposition") next to the feature(s) they produced in your profiles.

For each profile A through D, circle the number that represents the best sequence of events for that profile.

PROFILE A: **1.** Deposition, faulting, erosion, tilting, deposition

   2. Extrusion, faulting, metamorphism, melting, deposition

   3. Folding, erosion, deposition, tilting, faulting

   4. Deposition, folding, erosion, deposition, erosion

PROFILE B: **1.** Deposition, intrusion, erosion, contact metamorphism, deposition, erosion

   2. Deposition, erosion, deposition, extrusion, deposition, erosion

   3. Solidification, erosion, deposition, erosion, faulting, erosion

   4. Deposition, erosion, deposition, intrusion, deposition, faulting

PROFILE C: **1.** Deposition, metamorphism, melting, solidification, tilting, deposition, erosion, extrusion, erosion

   2. Metamorphism, melting, solidification, faulting, erosion, intrusion, extrusion, deposition, erosion

3. Deposition, metamorphism, melting, solidification, faulting, erosion, deposition, intrusion, erosion

4. Deposition, contact metamorphism, faulting, deposition, erosion, extrusion, deposition, evaporation, erosion

PROFILE D: 1. Solidification, deposition, tilting, erosion, deposition, intrusion, erosion, faulting, deposition, erosion, deposition, extrusion, erosion

2. Solidification, metamorphism, tilting, erosion, deposition, intrusion, erosion, deposition, faulting, erosion, deposition, extrusion, erosion

3. Solidification, deposition, tilting, erosion, deposition, intrusion, erosion, deposition, erosion, faulting, deposition, extrusion, erosion

4. Solidification, deposition, tilting, erosion, deposition, extrusion, erosion, deposition, faulting, erosion, deposition, extrusion, erosion

There are 10 different events mentioned in this lab: deposition, erosion, intrusion, extrusion, tilting, folding, faulting, metamorphism, melting, and solidification. (Use these words to answer the next three questions.)

1. Which events, from the list above, can we often observe around us?

　　_____

2. Which of these events (from the list above) changes one kind of rock directly into another kind of rock?

　　_____

3. Which of these 10 events destroy a part of the geologic record, such as fossils? (List at least two.)

　　_____

CHAPTER 17—LAB 4: A MODEL OF RADIOACTIVE DECAY

Introduction

The study of fossils has enabled geologists to establish a relative time scale of Earth's history. However, assigning numerical ages requires the use of complex laboratory procedures to measure tiny amounts of natural radioactive isotopes.

For example, carbon-14 is useful in finding the age of organic remains (such as wood, charcoal, or bone) from the very recent geological past. (Beyond about 50,000 years, so little carbon-14 is left that it is difficult to measure.) This lab will provide a conceptual model of radioactive decay at the atomic level. Ionizing radiation is invisible, but it can be measured with a Geiger-Müller counter, as shown in Figure 17-15.

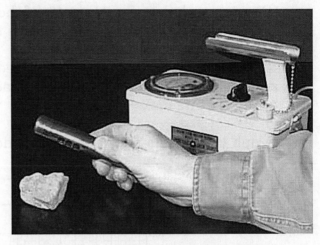

FIGURE 17-15. Using a Geiger-Müller counter to detect radiation.

Objective

To model radioactive decay.

Materials

50 small plastic disks (25 disks with one red side and one white side, 25 disks with two gray sides), plastic container with cover

Procedure

1. Count your disks. You must start with 25 disks with a white side and a red side. If you have more than 25, return the extras to their container. If you need more, your teacher can supply them. You will also need 25 disks with two gray sides.

2. Turn all 25 disks so that the white side is on the top. The white disks represent the original radioactive isotope. The red disks represent the most recent decay product. The gray disks will represent the older decay product.

3. Cover the container and shake it for about 5 to 10 seconds

4. Open the container and remove the disks that have turned to the red side. See Figure 17-16. Replace them with gray disks. Then count the number of white disks left and record this number in the data table in Figure 17-17. (Because each dot represents 4% of the total, the percentage is the number of disks times 4.)

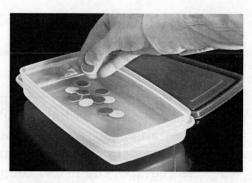

FIGURE 17-16.

5. Repeat steps 3 and 4, removing the red disks and replacing them with gray disks each time, until no white disks are left.

6. Graph your results in the place provided in Figure 17-17.

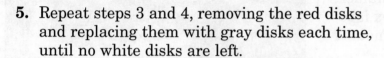

% of the 25 Disks Left

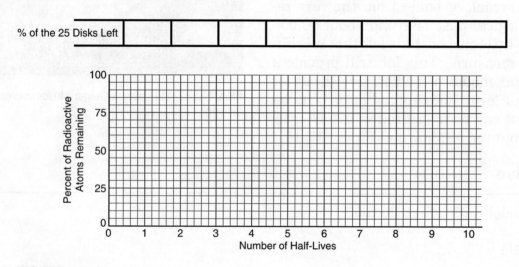

FIGURE 17-17.

Wrap-Up

1. What percent of the 25 original radioactive atoms are left after 1 half-life? _____ After 2 half-lives?

_____

2. Define half-life. _____

_____

3. Write a series of fractions to show the portion of the original radioisotope remaining after 1, 2, 3, 4, and 5 half-lives.

_____

4. Now write a series of fractions showing the portion of the decay product after 1, 2, 3, 4, and 5 half-lives.

_____

5. Write the name and half-life (as regular numbers, not in exponential notation) of the substances in the *Earth Science Reference Tables* that have the longest and shortest half-life.

_____

_____

6. What is the decay product of carbon-14? (Please give both the name and mass number of the element.)

_____

7. If you started with 10 grams of carbon-14, how much would remain as carbon-14 after 17,100 years?

_____

8. What radioactive substance has a half-life approximately as long as the age of Earth?

_____

9. Give two (2) reasons why carbon-14 could not be used to find the age of granite that is about a billion years old.

_____

_____

10. Graphs showing the radioisotope remaining through time always curve downward like the graph you made in this lab. Why can this graph line not be a straight line?

_____

_____

11. In what ways is the lab procedure unlike real radioactive decay?

_____

_____

_____

_____

12. Is the data you collected the same as data of other groups? What do the data of different groups have in common?

_____

_____

13. You may have learned that the decay of any radioactive atom is an unpredictable event. (This is like the probability of any one disk flipping over in your lab procedure.) If the decay of every atom is unpredictable, how can we use them to find a numerical age with a high degree of accuracy?

Chapter 18
Fossils and Geologic Time

 CHAPTER 18—LAB 1: A GEOLOGIC TIMELINE

Introduction

You may think that events that took place before you were born, or even centuries ago, are ancient. However, compared with the age of Earth, these events are extremely recent.

Objective

In this lab you will construct a time line by placing selected geologic events in the proper sequence. This model should help you gain a more realistic perspective of Earth's geologic history.

Materials

5-meter length of adding machine tape, meterstick, pencil, masking tape

Procedure

1. Obtain a 5-meter strip of paper tape. Neatly write the name of each member of your group on both ends of the strip of paper. See Figure 18-1.

2. Your timeline will be easier to construct (and to read) if you make a narrow reference line across the strip about 10 cm from one end. Neatly label this line "Present." All times will be measured from this line.

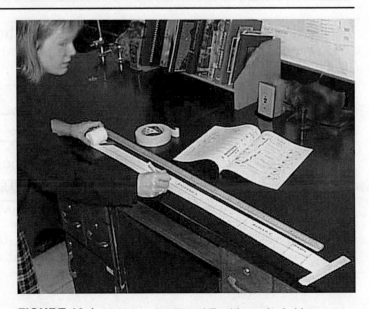

FIGURE 18-1. Making a timeline of Earth's geologic history.

3. Complete the middle column in the Table 18-1, below. You will fill in the last column after you set a scale.

TABLE 18-1. Timeline Data

Event	Age in Scientific Notation (years)	Age (years)	Scale Distance
Origin of Earth and Solar System	4.6×10^9	4,600,000,000	
Oldest rocks on Earth	4.2×10^9		
Earliest traces of marine organisms	3.9×10^9		
First animals with shells and skeletons	5.4×10^8		
Earliest fossils of land (terrestrial) animal	4.2×10^8		
First birds	1.6×10^8		
Extinction of dinosaurs	6.6×10^7		
Earliest humans evolve in Africa	3.0×10^6		
End of Wisconsin Glacial Era	1.0×10^4		
Columbus set sail	5.2×10^2		
First humans land on the moon (1969)	3.8×10^1		

4. Establish a scale that you will use to plot the events from the data table along your strip of paper. Answering the following questions will help.

   a. How old is our planet? _____

   b. How many billions of years is this? _____

   c. How long is your paper tape? _____

   d. To fit a little over 4.5 billion years on your paper and to use round numbers, it will be convenient to let each meter represent _____ billion years.

   e. How many centimeters are there in a meter? _____

   f. If each meter represents 1 billion years, how many years are represented by each centimeter? _____ (Best expressed in words)

5. Use the scale that you established in step 3 to complete the last column in the data table.

6. Plot the events from the data table in their proper place on your timeline. (It is easiest to start with the oldest events.)

7. Open your *Earth Science Reference Tables* to the geologic time scale, which provides information about Earth's geologic history. Note that the most recent portion has been expanded on the right side of the page. This is because rocks older than 540 million years seldom contain fossils. Does your timeline look like an expanded version of the line on the left side of the chart in the *Reference Tables*? _____

8. Divide your timeline into the four time epochs shown in the *Reference Tables*. Be sure that you have clearly shown when each one begins and ends. Label each according to the kind of life that was dominant.

   Precambrian: First Primitive Life Forms

   Paleozoic: First Organisms with Shells or Skeletons

   Mesozoic: Age of Dinosaurs

   Cenozoic: Age of Mammals

9. Use your completed timeline to help you answer the Wrap-Up questions.

10. When you finish, fold the timeline paper in half four times lengthwise (so that it will be about the length of this paper). Staple each group member's completed lab papers to the timeline and submit them.

Wrap-Up ▷

1. Which of these geologic ages lasted the longest? _____

   a. Precambrian   **b.** Paleozoic   **c.** Mesozoic   **d.** Cenozoic

2. For approximately what fraction of the Earth's history has there been life on our planet? _____

   a. $\frac{1}{4}$ **b.** $\frac{1}{2}$ **c.** $\frac{3}{4}$

3. According to the information in Table 18-1, in what environment did life on the Earth begin?

   _____

4. According to your timeline, what event marks the beginning of the Paleozoic Era?

   _____

5. What kind of life forms existed in the Precambrian?

   _____

6. What group of animals was dominant in the Mesozoic Era?

_____

7. On the timeline in Figure 18-2, clearly show each of the following events by writing its number at the proper position. (1) Origin of Earth, (2) Fossils become common (animals with hard parts), (3) Humans appeared, and (4) When I was born

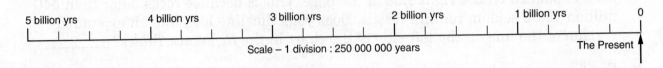

FIGURE 18-2.

8. Use a proportion to determine the length of a time line showing all the events in question 7 at a scale 1 cm:15 years (your approximate age). Clearly show the proportion and the steps to your solution. Your answer must be expressed in kilometers. (This question is not required, but it may be used to earn extra credit.)

CHAPTER 18—SKILL SHEET 1: READING GEOLOGIC CHARTS

A useful phrase for students of Earth science is, "If in doubt, look it up in the *Reference Tables.*" You may be surprised at the amount of useful information in that document. The following questions can be answered using the *Earth Science Reference Tables* (ESRT). Most answers will be found on pages 2, 3, and 8–9 of the ESRT.

1. What is the metric distance from Elmira to Ithaca? _____

2. What is the metric distance from Buffalo to Albany? _____

3. Compared with a mile, approximately how long is a kilometer? _____

4. What is the (relative) geologic age of the bedrock near Syracuse, New York?

_____

5. What is the numerical age of the bedrock at 43°30′N, 76°00′W?

_____

(You will need to use page 3 and pages 8–9.)

6. New York State rocks of Devonian age are mostly _____

    a. Igneous   **b.** Sedimentary   **c.** Metamorphic

7. Where is the youngest "bedrock" in New York? [Actually, it is mostly unconsolidated (soft) sediments at the surface.]

_____

8. What geographic feature is found in the New York State region that has the oldest bedrock?

_____

9. Bedrock of what geologic age underlies most of the Allegheny Plateau?

_____

10. Most of New York is covered by rocks that are _____

    a. Igneous   **b.** Sedimentary   **c.** Metamorphic

11. Where you live, most of the bedrock is _____

    a. Igneous   **b.** Sedimentary   **c.** Metamorphic

12. Near what city is the bedrock of Pennsylvanian and Mississippian age?

_____

13. You have decided to go into the rock supply business. If you wanted to quarry marble, in what two parts of New York State could you locate your quarry?

_____

14. What small New York town is surrounded by a large area of bedrock that is unlikely to contain fossils?

_____

15. The Apollo Program brought back anorthosite rocks from the moon. Where is anorthositic rock found in New York State?

_____

16. As the Genesee River flows toward Lake Ontario, it flows through rocks that are progressively _____

    a. Younger   **b.** Older

17. What line of latitude extends along large portions of New York's southern boundary?

_____

18. New York State has few igneous rocks. However, the most impressive outcropping forms the late Triassic Palisades Sill. What are the approximate latitude and longitude of this feature? _____

19. Why are geologists unsure of the age of the bedrock at 43°30′N, 77°30′W?

_____

20. Why is the top part of the timeline on the left of page 8 of the ESRT expanded toward the right?

_____

_____

_____

21. When did the Mesozoic Era begin and end. (Give the beginning time first.)

_____

22. How long ago did Earth consolidate from a cloud of debris in space?

_____

23. What ancient orogeny formed the Adirondack Mountains and the Hudson Highlands? (An orogeny is a period of mountain building.)

_____

24. Which of these four periods of Earth history is longer than all other three combined? _____

    a. Cenozoic   **b.** Mesozoic   **c.** Paleozoic   **d.** Proterozoic

25. Which biological life forms evolved first? _____

    a. Birds   **b.** Dinosaurs   **c.** Flowering plants

26. Which epoch of geologic time has been most brief? _____

27. What geologic process destroyed most of New York's geologic record of the Cenozoic Era?

_____

28. What geologic period is not represented in the bedrock of New York State?

_____

29. Between the Triassic and the Cretaceous periods, which continent moved away from Africa? (See the maps on the right side of ESRT page 9.)

_____

30. Near what city would a geologist be most likely to find dinosaur fossils? (Actually, dinosaur footprints have been found there.)

_____

31. The Precambrian forms approximately what portion of Earth's total history?

_____

32. What four geologic periods are completely represented in New York's bedrock?

_____

33. Early humans could have been hunted what two animals, which are now extinct?

_____

34. What group of animals evolved early in the Paleozoic and became extinct near the end of the Paleozoic?

_____

35. Only one kind of dinosaur is known to have left fossils in New York State. What dinosaur was it?

_____

36. The Catskill Mountains are unusual in that the bedrock is mostly flat-lying sedimentary layers. In what kind of landform were these sediments deposited in the late Devonian Period?

_____

37. Where was New York State (or at least the bedrock that would become New York State) located during the Ordovician Period?

_____

38. How old are the oldest known rocks in the bedrock of New York State?

_____

39. Most of the animal organisms shown at the top of pages 8 and 9 of the ESRT are invertebrates. Name one exception.

_____

40. What major geologic event or feature follows both episodes of rifting in New York's tectonic history? (This two-word term refers to a boundary that is relatively inactive in the processes of mountain building.)

_____

41. The fossil in Figure 18-5 was found in a New York State bedrock outcrop. What is the approximate latitude of the most likely area where this fossil was found?

_____

_____

_____

_____

FIGURE 18-5. The Eurypterid (sea scorpion) was adopted as the New York State's fossil in 1984.

Copyright © 2007 AMSCO School Publications, Inc.

Chapter 19
Weather and Heating of the Atmosphere

 CHAPTER 19—SKILL SHEET 1: HEAT FLOW

Heat can travel, or flow, from one place to another in three ways: by convection, conduction, and radiation. Heat moves through nontransparent solids primarily by conduction. If you could see atoms and molecules at any temperature above absolute zero, you would see them vibrate with heat energy. Absolute zero (0 K or −273°C), like the speed of light, is one of the few known absolute limits in our universe. It is not possible to have a temperature below 0 K because at that temperature, molecular motion stops. The hotter a substance is, the more its molecules vibrate.

By progressive collisions of molecules, hotter molecules, which vibrate more, pass along their energy to cooler neighbors, increasing their vibration. Heat energy always flows from something that is hot (a heat source) to something that is cold (a heat sink). In Figure 19-1 heat is traveling up through the metal spoon as hot molecules increase the vibrations of nearby cooler molecules.

FIGURE 19-1. Heat flows through the metal spoon by conduction.

1. How does heat flow in the process of conduction?

_____

_____

2. *a.* Name one solid that is a good conductor of heat energy.

_____

** *b.*** Name one solid that is a poor conductor of heat energy.

_____

Unlike by conduction, when heat flows by convection the energy is carried by atoms and molecules as the medium (heated substance) moves. Convection is caused by differences in density within a fluid (usually a liquid or a gas). Figure 19-2 illustrates convection in the ocean. The flag in Figure 19-3 is being moved by the wind, which is also caused by convection.

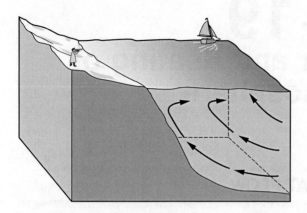

FIGURE 19-2. Convection distributes heat energy within the oceans.

FIGURE 19-3. Wind is convection within the atmosphere.

Circulation by convection generally forms a closed circuit. For example, when a stove heats a room, warm air near the stove rises because warm air is less dense than cooler air. As the warm air moves along the ceiling to the cool walls, it becomes cooler and more dense, sinking to the floor. Circulation along the floor back to the stove completes the convection cell. At that point, air is heated by the stove and rises again.

3. What usually causes a heated fluid to rise?

_____

4. In Chapter 8 you learned about circulation within Earth's mantle even though it is generally considered a solid. What global system of change in Earth's lithosphere is driven by these convection currents?

_____

Light and heat can travel through empty space by radiation. Figure 19-4 shows light from the sun reaching Earth. Radiation is the only form of heat flow that requires no medium (substance) through which to pass. Radiation can also carry heat energy through transparent solids and liquids. Radiation travels at 3×10^8 meters per second, so it moves extremely fast. Like absolute zero, the speed of electromagnetic radiation is also a physical limit within our universe.

FIGURE 19-4. Radiation is the only form of heat flow through a vacuum.

5. How does energy travel from the stars to Earth? _____

6. Light, heat (infrared), and ultraviolet are all forms of what kind of energy?

_____

For questions 7–11, indicate the form of heat flow described: conduction, convection, or radiation.

7. Heat flow by density currents in fluids _____

8. Rapid heat flow through glass or clear water _____

9. Energy transfer through opaque solids (especially metals) _____

10. How Earth loses energy to space _____

11. Energy passed from molecule to molecule _____

For questions 12–19, indicate the form of heat flow being illustrated.

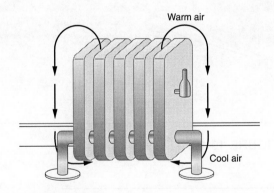

FIGURE 19-5.

FIGURE 19-6.

12. _____   **13.** _____

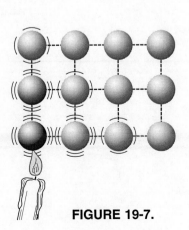

FIGURE 19-7.

FIGURE 19-8.

14. _____   **15.** _____

FIGURE 19-9.

FIGURE 19-10.

16. _____

17. _____

FIGURE 19-11.

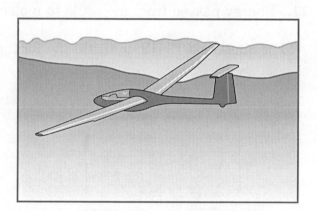

FIGURE 19-12.

18. _____

19. _____

20. Which form of heat flow requires no medium through which to pass? _____

CHAPTER 19—SKILL SHEET 2: INSOLATION

Earth receives almost all of its energy from the sun. Compared with billions of other stars, the sun is average in size, color, and temperature. Stars produce energy by the process of nuclear fusion. Deep inside stars, four hydrogen atoms combine to become a single atom of helium. During this process, a small amount of matter is converted to a vast amount of energy.

Solar energy rises to the sun's surface by convection. This heats the surface of the sun to about 6000°C. From the surface, the energy is radiated outward as heat, light, and other forms of electromagnetic radiation, most of which travels out of the solar system into deep space. Only a tiny fraction of this energy reaches the planets. Sunlight that reaches Earth takes only 8 minutes to travel about 150 million km. Radiation is the only form of energy transfer that can take place through space.

The solar radiation that reaches Earth is known as insolation. Insolation is a contraction of three words, INcoming SOLar radiATION. Although sunlight contains a wide spectrum of wavelengths from heat rays to X-rays, solar radiation is strongest in the wavelengths of visible light.

1. What single word means the same as insolation? _____

2. What happens to most of the sun's radiant energy?

_____

3. What two forms of heat flow carry energy from inside the sun to Earth?

_____

4. In nuclear reactions, what is lost as energy is created? _____

Insolation interacts with Earth in several ways. Have you ever viewed a pencil or spoon in a glass of water from the side, noticing that it seems to disconnect at the surface? The bending of light rays as they pass from the water to the air causes the apparent bending of the spoon or pencil. This bending of light is known as refraction. As low-angle rays of sunlight penetrate increasingly dense layers of the atmosphere, the light slows causing it to refract, or bend, downward toward the ground. Figure 19-13 shows refraction and three other ways that light behaves when it reaches Earth.

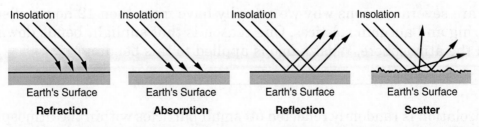

FIGURE 19-13. Insolation interacts with the atmosphere and Earth's surface in four ways.

If the ground were a mirror, all the light would bounce off like balls hitting a smooth floor. When light bounces off a surface scientists call it reflection. If the surface is rough, and the direction of reflection is random, the light scatters.

Light that is not reflected is absorbed. Most insolation that is absorbed changes to heat energy and warms the surface. Light-colored surfaces reflect more energy, and dark-colored surfaces absorb more energy. Table 19-1 shows the ability of different natural surface materials to reflect and absorb insolation.

TABLE 19-1. Absorption and Reflection

Material	Absorption (%)	Reflection (%)
Snow	25	75
Dry Sand	75	25
Grass	85	15
Water	60–90	40–10
Forest	95	5

Have you noticed that when you look straight down into a clean, still pool of water, you can see into the water? However, if you look at a low angle, the smooth water surface acts like a mirror. Therefore the albedo (reflectivity) of some surfaces is variable, depending on the angle of the sun. In general, more sunlight is reflected when the surface is light in color and smooth and the sun is low in the sky.

5. According to Table 19-1, what ground cover absorbs the most insolation?

_____

6. Give two reasons why, compared with the tropics, relatively little sunlight is absorbed in Earth's Polar Regions.

_____

_____

7. What happens to most of the energy when Earth's surface absorbs light?

_____

8. There are several reasons why we actually have more then 12 hours of sunlight on the spring and autumn solstices. One reason is that sunlight bends downward within the atmosphere. What name is applied to this bending?

_____

Some insolation is randomly reflected off small particles within the atmosphere. These particles are called aerosols. Aerosol particles can be pollution from cars, homes, and industries; dust; salt spray; cloud droplets; or ice crystals. These particles remain suspended because they are too small to fall through the atmosphere. Figure 19-14 shows how aerosols affect insolation.

When salt water becomes too concentrated with salt, the salt settles out, or precipitates. Similarly, when the atmosphere becomes saturated with moisture, water can fall to the ground as rain or snow. This is also called precipitation.

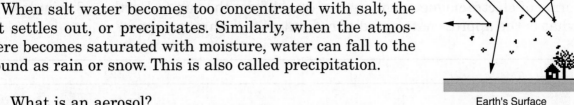

FIGURE 19-14. Random reflection off aerosols scatters insolation within the atmosphere.

9. What is an aerosol? _____

_____

10. How do aerosol particles affect insolation?

_____

11. What name is applied to the settling-out of substances in a solution and to rain or snow?

_____

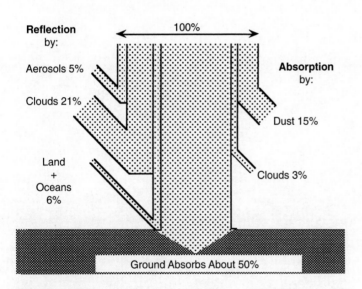

FIGURE 19-15. Earth reflects and absorbs insolation.

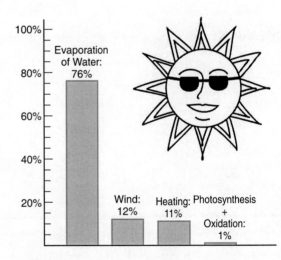

FIGURE 19-16. Earth uses solar energy.

12. According to Figure 19-15, the ground absorbs the largest percentage of energy, what happens to the next largest percent?

_____

13. Figure 19-16 shows the changes insolation causes on Earth. What is the primary change caused by solar energy?

_____

14. Insolation is a combination of what three words?

_____

15. Light is just one form of electromagnetic energy. Other wavelengths of electromagnetic energy include radio waves, heat, ultraviolet, and X-rays. If these forms of electromagnetic energy can travel from the sun to Earth in 8 minutes, what is the approximate speed of radiant energy through space?

_____

_____

CHAPTER 19—LAB 1: ANGLE OF INSOLATION

Introduction

The seasons are caused by the annual cycle of the changing angle of the sun in the sky. If the seasons were caused by changes in the distance to the sun, we would have summer in January, when Earth is closest to the sun in its orbit. This lab is a model of insolation (sunlight) striking the ground at three different latitudes on Earth. The 90° thermometer represents a place at the equator and the 30° thermometer a location near the poles.

Objective

In this lab you will investigate how the angle at which the light strikes a surface affects the heat energy absorbed by the surface.

Materials

3 metal-backed thermometers, angle template or protractor, black thermometer-bulb covers, metal lamp (100 watts), books (to elevate thermometers), metric ruler

Procedure

1. Select three thermometers that read within 1°C of one another. Replace thermometers that do not agree with others that do agree.

2. Slide each thermometer bulb inside a small black paper envelope. Place the thermometers next to one another in a holder. Set this apparatus about 20 cm from a light source as shown in Figure 19-17. It is very important that the center of the thermometer bulbs and the light bulb be at the same height above the desk or lab surface.

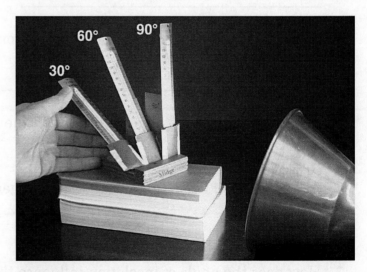

FIGURE 19-17. Thermometers set at a 30°, 60°, and 90° angle to light rays.

3. Record the initial temperatures below, then turn on the lamp and record the three temperatures each minute for a total of 15 minutes.

TABLE 19-2. Angle and Temperature Data Table

Time (min)	0														
Temp. at 30°															
Temp. at 60°															
Temp. at 90°															

4. Graph your data. You will make three (3) best-fit lines on a single graph. Label them 30°, 60°, and 90°. (Your teacher can show you how to make a "best-fit line.")

Wrap-Up

1. The thermometer at which angle became hottest? _____

2. Which thermometer angle represents a place where the sun is straight overhead?

_____

3. At the time of the equinoxes, where is the sun straight overhead at its highest point in the sky?

_____

4. At what time of day is the angle of insolation the greatest? (That is, at what time is the sun highest in the sky?)

_____

5. In what month is the angle of insolation the greatest in the United States?

_____

6. As the angle of insolation increases, what generally happens to the surface temperature?

_____

7. How does the angle of insolation influence the cycle of Earth's seasons?

_____

_____

_____

8. What is the relationship between the angle of the thermometers and the latitude of the places they represent on Earth?

_____

_____

Figure 19-18 shows that when the sun is high in the sky, rays of sunlight are concentrated into a small area on Earth's surface. When the sun is low, the same insolation is spread over a large area and insolation cannot heat the Earth as quickly.

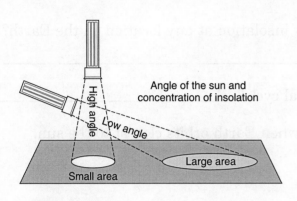

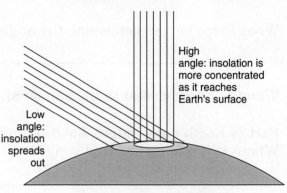

FIGURE 19-18.

9. As the angle of insolation increases, what happens to the intensity of insolation at Earth's surface?

_____

10. Why is sunlight always weak at Earth's poles? _____

_____

11. As latitude increases, how does the angle of insolation generally change?

_____

Figure 19-19 is a three-dimensional graph that shows the strength of insolation over the whole Earth during one year.

12. What quantities are shown on the three axes of this graph? (The vertical axis is not labeled but the quantity should be clear.)

_____

_____

_____

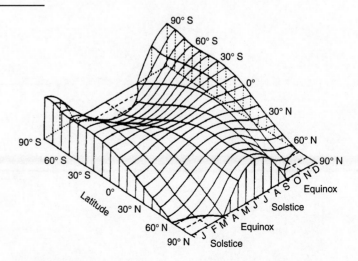

FIGURE 19-19. The annual cycle of insolation energy over the whole Earth.

13. In what month does the North Pole receive the most intense insolation?

_____

14. Where on Earth is the strength of insolation nearly constant throughout the year?

_____

15. For how many months does the North Pole receive no direct insolation?

_____

16. What three factors determine the angle of insolation at any location on the Earth?

_____

17. When is Earth closest to the sun on annual cycle? _____

18. Part of Earth does, in fact, have summer when Earth orbits closest to the sun. Where is it summer at that time?

_____

🌐 CHAPTER 19—LAB 2: CONDUCTION

Introduction

Heat energy travels by conduction, convection, and radiation. Conduction can occur within solids, liquids, and gases. Conduction is the only form of heat energy transfer through opaque solids. Conduction occurs when heat energy (vibrations of atoms and molecules) excites nearby atoms and molecules to vibrate with increasing energy. Insulators are materials that resist conduction.

An energy source is an object that has more energy than its surroundings. Usually, an energy source is relatively warm. An energy sink is anything that is cooler, or lower in energy than its surroundings. The net flow of heat energy is always from an energy source to an energy sink.

Objective

To investigate heat flow by conduction.

Materials

2 foam calorimeters, 2 calorimeter covers, 2 Celsius thermometers, metal conduction bar, stopwatch or a clock with a second hand

Procedure

1. Put cold tap water in both foam calorimeters to be sure they do not leak. If a container leaks, tell your teacher.

2. Fill one calorimeter $\frac{3}{4}$ full of hot tap water. Fill the second $\frac{3}{4}$ full of cold tap water.

3. Push the aluminum bar through the two calorimeter covers, then insert the two thermometers as shown in Figure 19-20. Be sure the thermometer bulbs are below the surface of the water.

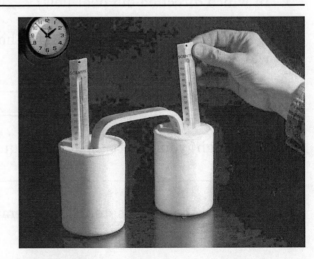

FIGURE 19-20. Heat flows through the aluminum bar by conduction.

4. In Table 19-3, record the initial Celsius temperature and then the temperature each minute for 20 minutes.

TABLE 19-3. Conduction Data Table

Time (min)	0	1	2	3	4	5	6	7	8	9	10	11	12	13	14	15	16	17	18	19	20
Hot Water																					
Cold Water																					

While you are collecting data you can work on your graph. Place time on the horizontal axis. Draw two best-fit lines (gentle curves) on a single graph. If you need help making best-fit limes, your teacher can help you. Be sure both axes are properly labeled with quantities and units.

The hot calorimeter changed by _____° in 20 minutes.

The cold calorimeter changed by _____° in 20 minutes.

Wrap-Up

1. Which calorimeter was the heat source? _____

2. Which calorimeter was the heat sink? _____

3. In which direction did heat flow? _____

4. Did the energy lost by water in the hot calorimeter equal the energy gained by water in the cold one?

   _____

   _____

5. Why might you expect them to be equal?

   _____

6. When did the calorimeter change temperature at the fastest rate? (Look at your graph.)

   _____

7. Why did the rate of heat flow decrease with time? _____

   _____

8. Where else could the energy from the hot calorimeter have gone besides into the cold water?

   _____

CHAPTER 19—LAB 3: ABSORPTION AND RADIATION

Introduction

Why do you wear light colors in the summer and dark colors in the winter? To help you understand why, you will be recording the temperature of the two cans as they absorb and then loose heat. (You may be surprised by part of the results.)

Objective

In this experiment you will see how an object's color affects its ability to absorb visible radiation and then to give off invisible heat radiation.

Materials

1 black can, 1 silver can, lamp with reflector, insulating covers for cans, books to elevate the cans, 2 thermometers, ruler or meterstick

Procedure

1. Find two thermometers that are displaying the same Celsius temperature. Set up the apparatus as shown in Figure 19-21. The two cans should be about 2 to 4 cm apart. Be sure that the center of each can is at the same height as the center of the light bulb, and that each can is exactly 15 cm from the light bulb. ***Do not turn the light on yet.***

2. In Table 19-4, record the initial temperature of both cans in the 0 min boxes below. (Ask your teacher to check your setup before you turn on the light.)

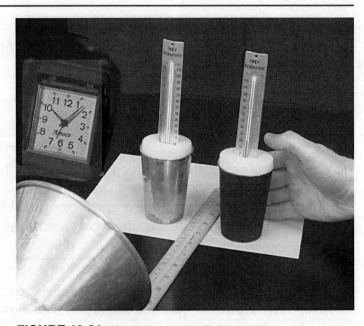

FIGURE 19-21. Absorption and radiation setup.

3. When you have your teacher's approval, turn on the light. Record the temperature of each can each minute for 10 minutes while the lamp is on.

TABLE 19-4. Absorption Data Table

Time (min)	0	1	2	3	4	5	6	7	8	9	10		11	12	13	14	15	16	17	18	19	20
Black												Turn Off the Lamp										
Silver																						

4. After exactly 10 minutes, without disturbing the cans, turn off the lamp and slide it away. Continue recording the temperatures for another 10 minutes as the cans cool.

5. Construct a graph of your data. This graph will have two best-fit lines on one graph: one line for the silver can and one for the black can. You can start your graph while you are recording the last 10 minutes of data.

Wrap-Up ▌▌▌➤

1. Which color (black or silver) appears to absorb electromagnetic energy better?

_____

2. According to your graph, which color cools faster? (That is, which graph line has a steeper downward curve?)

_____

3. On a cool, sunny day, what color clothing would keep you the warmest? _____

4. What color clothing will keep you cool on a hot, sunny summer day? _____

5. What color is best to keep a car a comfortable temperature all year long? This color should help the car stay cool in summer and hold its warmth in the winter.

_____

You may know that Earth receives nearly all its energy as insolation. However, Earth also radiates energy back into space. If energy gained and energy lost are equal, Earth's average temperature remains stable.

For the most part, you cannot observe the radiant energy that Earth gives off because it is invisible infrared heat radiation. Sometimes you can tell a surface is hot and giving off heat radiation by placing your hand near it. (See Figure 19-22.) Your skin can detect strong heat radiation.

FIGURE 19-22. Not all radiant energy is visible to us.

6. Is heat radiation visible to our eyes? _____

7. How can the person in Figure 19-22 tell that the iron is radiating energy?

_____

Earth absorbs a variety of wavelengths of electromagnetic energy, but the most intense part of the radiation is in the visible light wavelengths. The energy radiated by Earth is mostly long-wave, heat energy also known as infrared radiation.

8. According to Figure 19-23, what happens to the wavelength of energy when it is absorbed by Earth and then radiated as infrared energy?

_____

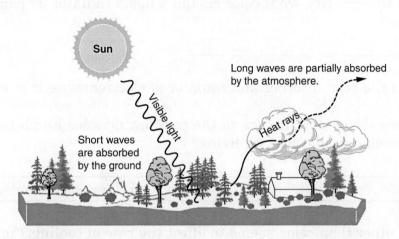

FIGURE 19-23. The ground absorbs light rays and radiates infrared energy.

Figure 19-24 shows light energy and other forms of radiation that reach Earth's surface. It also shows the quantity of infrared (heat) energy that Earth radiates from its surface. Notice that most of the dangerous ultraviolet radiation from the sun (the white band on the left) is absorbed by our atmosphere. Ozone is the principal gas that protects us from this dangerous short-wave radiation.

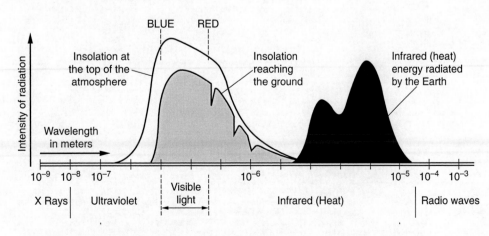

FIGURE 19-24. Earth's average temperature depends on the balance between insolation and terrestrial radiation.

9. What would happen to Earth if it did not give off as much energy as it absorbs?

_____

10. According Figure 19-24, in what part of the electromagnetic spectrum is the solar radiation that reaches the ground most intense?

_____

11. In what part of the spectrum is the terrestrial radiation most intense?

_____

12. According to this activity, what color should a home radiator be painted to give off the most heat?

_____

13. What color is the best absorber and radiator of electromagnetic energy? _____

14. What evidence shows that, in fact, in the past few decades Earth has been radiating less energy that it is receiving?

_____

_____

15. What factor, other than color, seems to affect the rate of cooling. (Your graph should show the cooling line for both cans become less steep as time passes. Why does this happen?)

_____

_____

CHAPTER 19—LAB 4: INQUIRY INTO ABSORPTION AND RADIATION

Introduction

In this lab you are to design and conduct a procedure to help you understand how different colors and textures absorb heat energy.

Objective

Which ice cube will melt fastest, one wrapped in something shiny or one wrapped in something dark?

Materials

Present your list of materials to your teacher for approval.

Procedure

Devise a procedure and submit it to your teacher for approval.

Write a lab report in which you give your objective, materials, procedure, and conclusions.

My Notes

Chapter 20
Humidity, Clouds, and Atmospheric Energy

☀ CHAPTER 20—SKILL SHEET 1: THE KINETIC THEORY OF MATTER

Energy is the ability to do work. The many forms of energy include chemical, mechanical, electrical, and atomic. Energy in storage is known as potential energy. A book on a shelf has potential energy because it can freely fall to the floor and do work, such as making a sound. However, energy in motion is kinetic energy. While the book is falling, it has kinetic energy. The wind has kinetic energy, which can be harnessed to pump water or make electricity.

The most important form of energy in Earth processes is heat energy. Heat can be kinetic energy and potential energy.

All matter is composed of atoms and molecules. Although these particles are too small to see, above a temperature of absolute zero (−273°C or −470°F) they are in constant vibration as illustrated in Figure 20-1. That vibrational energy involves motion so it is kinetic heat energy. When you measure temperature you measure the average vibrational energy of molecules.

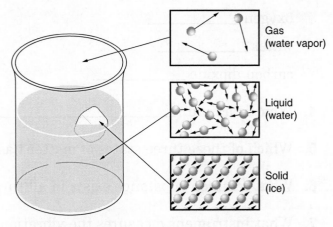

FIGURE 20-1. Molecules are in constant motion.

As the temperature cools toward absolute zero, that vibration slows. Theoretically, at absolute zero (0 K) all molecular motion would stop. Like the speed of light, absolute zero is a physical limit that can be approached experimentally, but never quite reached. However, using special laboratory techniques, scientists have been able to reach temperatures within a fraction of a degree of absolute zero.

1. What is the Celsius temperature of absolute zero? _____

2. What happens to atoms at a temperature of absolute zero?

_____

Heat also exists as potential energy. Potential heat energy is known as latent (or hidden) energy. The amount of latent potential energy in a substance determines whether that substance exists as a solid, a liquid, or a gas.

Solids contain the least potential energy. The atoms within a solid occupy fixed positions. Although they can vibrate with kinetic heat energy. Therefore solids have a relatively fixed size and shape.

Liquids contain molecules that can slide over and around each other like vibrating Ping-Pong balls in a bucket. Liquids have a relatively constant volume, but they take the shape of their container.

Gases have the most potential energy. The molecules in a gas have so much energy they bounce around in all directions, limited only by the walls of their container. Neither the shape nor the volume of a gas is fixed.

Liquids and gases can flow, for example water in a river or the wind. Therefore they are both fluids. As you learned in Chapter 8, Plate Tectonics, even what we think of as solid rock within Earth over long periods of time can flow.

The idea that matter is composed of tiny particles called atoms or molecules, that these particles are in constant vibrational motion, and that matter exists in several states is the kinetic theory of matter.

3. Label the following as solid, liquid, or gas according to how they are usually found in our terrestrial (Earth) environment.

    oxygen _____   iron _____

    water _____   mercury _____

    carbon dioxide _____ rock _____

4. What is a fluid? _____

5. Which of these three states of matter has the most latent energy? _____

6. What common substance exists in all three states on Earth? _____

7. What instrument measures the vibrational kinetic heat energy of molecules?

    _____

8. A stream can be though of as an energy conversion system. Describe the changes in kinetic and potential energy that occur as water flows from the stream's source to its mouth. If you can, sketch a graph of each.

    _____

    _____

    _____

    _____

CHAPTER 20—LAB 1: CHANGES IN STATE

Introduction

Changes in state play an important part in our weather because, during changes in state, large amounts of energy are absorbed and released. You have learned that when matter is heated, its temperature may rise because the atoms and molecules vibrate more vigorously. This is kinetic heat energy. However, a change in state is a change in potential heat energy. This potential heat energy is known as latent energy because it does not cause a temperature change.

Objective

To observe the effect of heating ice and water.

Materials

Safety goggles, 250-mL beaker, hot plate or gas burner, thermometer, stirring rod, timer or clock, snow or crushed ice, mass scale, graph paper (If you use a gas burner, you also will need matches, ring stand, and wire gauze.)

Precautions

1. Use the stirring rod to *stir the mix almost constantly* until the ice has melted. Do not use the thermometer or rest it on the bottom of the beaker.

2. Keep your beaker heating at a constant rate. Do not change the setting of heat dial or flame.

3. Record the temperature when the last of the ice has melted, and when the water first comes to a full boil.

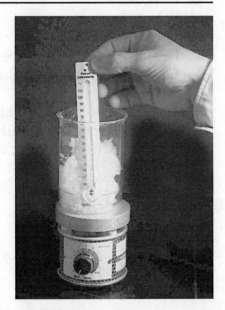

FIGURE 20-2. Observing kinetic and latent heat energy.

4. You **must** wear eye protection whenever you are within 2 meters (6 feet) of a flame or a liquid on a hot burner.

Procedure

1. If you are using a hot plate, preheat it on the highest setting for about 5 minutes. Position the hot plate where no person or object is likely to touch it.

2. Determine the mass of a 250-mL beaker. Record the mass on your data sheet. Add 200 mL of crushed ice or snow. Record the mass again. Determine the mass of the ice.

3. Carefully add about 1–2 cm of cold water to the beaker. Measure and record the mass of the ice-water mixture. Determine the mass of the water.

4. Predict the temperature of the ice-water mixture. _____°C (*Hint:* It is the temperature at which ice melts, and water freezes.) Rest the thermometer in the ice-water mixture until the reading is stable and record the temperature on your data sheet as the initial temperature.

5. Before you continue, reread the precautions. If you do not follow them carefully, you may need to start the procedure over again.

In the data table, record the Celsius temperature of the beaker each half-minute as you heat it. Keep the heating constant and stir the water with the stirring rod at least until all the ice has melted. Continue heating until the water has been boiling vigorously for a full 5 minutes. (Do ***not*** let the water boil away completely.) After 5 minutes of a full boil, turn off the heat.

6. Determine how much water has been changed to water vapor. (Your data table will help.)

7. Graph your temperature data. Be sure you have time on the horizontal axis. Do not forget the units of measure on both axes.

8. Clearly label the point along the graph line at which all the ice melted, and when the water comes to a full, rolling boil. Your graph should look like the part of this graph in the dotted box in Figure 20-3.

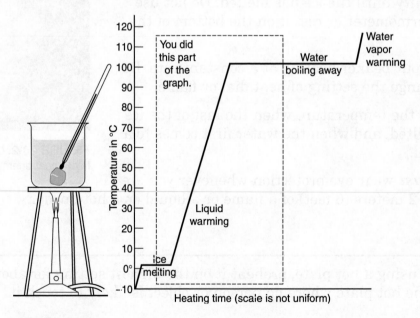

FIGURE 20-3.

Note that as energy is added, the temperature of the water does not rise constantly. Some of the time, as energy is added, the temperature of the water increased. In the flat sections of the graph energy was absorbed for a change in state. Melting and boiling absorb energy without any change in temperature. Therefore, latent energy is potential energy that is lost or gained in a change in state. Latent means hidden. This energy is hidden because it does not show as a change in temperature.

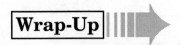

1. Which has more energy, water at 100°C or water vapor at 100°C? Why?

_____

_____

2. What is latent heat? _____

_____

_____

3. *a.* Which state of matter has the most latent energy? _____

   b. Which has the least? _____

4. Which of the lines on the graph in Figure 20-4 best shows the change in temperature of an ice-water mixture that was heated to boiling and then heated a few more minutes?

_____

5. Why does the temperature not rise above 100°C?

_____

_____

6. In your experiment, why was there no change in temperature during some of the time as energy was added?

_____

_____

7. According to Figure 20-5, which state of matter has the most latent energy?

_____

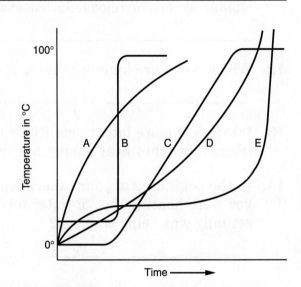

FIGURE 20-4.

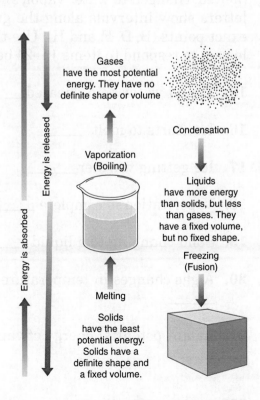

FIGURE 20-5. A change in state is a change in latent, potential energy.

8. Name two changes in state that require the addition of latent potential energy.
    _____ and _____

9. Name two changes in state that release latent potential energy.
    _____ and _____

10. A change in temperature is a change in what kind of heat energy.
    _____

11. When a cloud forms, water vapor changes to tiny droplets or ice crystals. Does this change absorb or release latent energy?
    _____

12. Which has more kinetic heat energy, ice at 0°C or water at 0°C?
    _____

13. Which has more latent potential energy, ice at 0°C or water at 0°C? _____ (Remember that heat energy has two forms.)

14. In the beginning of your experiment, there was very little temperature change as you were heating the ice-water mixture. If the temperature was constant, what actually was being changed?
    _____

    Figure 20-6 shows the change in temperature of a block of ice that was heated until it melted, changed to water vapor, and is then heated to 120°C. Please note that some of the letters show intervals along the graph (A, C, E, G, and I), while other letters represent exact points (B, D, F, and H). Use these letters to respond to items 15–24 below.

15. Liquid water being heated _____

16. Ice starts to melt _____

17. Ice getting warmer _____

18. Vaporization is complete _____

19. Solid changing to a liquid _____

20. A gas changes in temperature
    _____

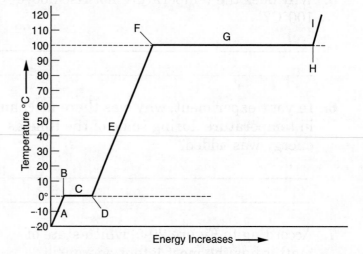

FIGURE 20-6. Heating curve of water.

21. Latent potential energy of vaporization is being added to the liquid _____

22. Molecules are being dislodged from their fixed positions in an ordered solid _____

23. Kinetic energy is increasing _____, _____, and _____

24. Latent potential energy is being added _____ and _____

Respond to items 25–32 by writing **A, B,** or **Both** according to which diagram in Figure 20-7 is described.

25. Changing temperature _____

26. Change in phase _____

27. Receiving more energy _____

28. No change in state _____

29. Fusion energy added _____

30. Kinetic energy added _____

31. No temperature change _____

32. Latent heat increasing _____

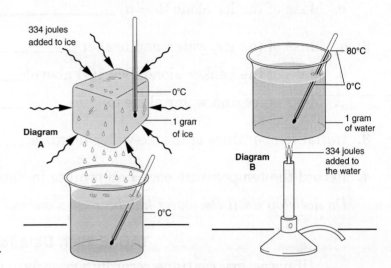

FIGURE 20-7. Diagram A and diagram B

33. What event carries large quantities of energy into Earth's atmosphere?

_____

_____

34. Clearly, the processes of melting and evaporation of water absorb energy (334 and 2260 joules per gram, respectively). What does this energy do?

_____

_____

35. How do refrigerators and air conditioners work?

_____

_____

_____

Data Table: Changes in State

Suggestions

1. Do this lab with a partner. One person can read the directions and record data, while the other partner performs the experiment.

2. Do not use this data table to guide you in doing lab. Use the lab paper instead.

3. Follow all directions in the lab sheets. Be sure to supply the needed metric units.

4. Do not forget that you must make a graph. Submit one lab, one data sheet, and one graph per person. Work neatly. If you need help, please ask.

1. **a.** Mass of the beaker alone (Step 2 in the lab procedure) _____

   b. Mass of the ice and the beaker _____

   c. Mass of the ice alone ($b - a$) _____

2. **a.** Mass of the ice, water, and beaker _____

   b. Mass of the beaker alone [See 1(a) above] _____

   c. Mass of ice and water alone ($a - b$) _____

3. Initial temperature of the ice-water mixture _____

4. Record the temperature each half minute in Table 20-1.

   Do not stop until the water has been at a vigorous boil for 5 minutes.

TABLE 20-1. Data Table
(If necessary, continue recording your data on another sheet of paper.)

Time	0 min	½ min											
Temp.													

Time													
Temp.													

   a. The ice was completely melted after _____ minutes.

   b. The water began to boil vigorously after _____ minutes.

5. **a.** Mass of the beaker and water at the end of the experiment _____

   b. Mass of the beaker, water, and ice at the start [2(a) above] _____

   c. Mass of water lost ($b - a$) _____

CHAPTER 20—SKILL SHEET 2: DEW POINT AND RELATIVE HUMIDITY

1. Define dew point.

   _____

   _____

2. What is the only way to change the dew point?

   _____

Determining Dew-Point Temperature and Relative Humidity

3. Tables used to determine the dew point and relative humidity are in the *Reference Tables* on page _____ .

4. To find the dew point subtract the _____bulb temperature from the _____ bulb temperature. The result of this subtraction is the _____.

5. On the dew-point temperature table follow the horizontal row across from the _____ temperature until it meets the vertical column coming down from the _____.

6. At the intersection of the row and column, read the _____ temperature.

7. The Relative Humidity table works the same way, except that instead of reading the dew-point temperature, you will read the _____.

8. | Dry-Bulb | | Wet-Bulb | Dew point: |
   |---|---|---|---|
   | *a.* | 20°C | 17°C | _____ |
   | *b.* | 10° | 1° | _____ |
   | *c.* | 5° | 3° | _____ (Interpolate!) |
   | *d.* | 8° | 8° | _____ |

9. What do you mean when you say that the humidity has reached saturation?

   _____

   _____

10. What term indicates the moisture content of the atmosphere measured as a percentage of saturation?

   _____

11. Dry Bulb   Wet Bulb   Relative Humidity

    a. −20°C   −21°C   _____

    b.   12°   10°   _____

    c.   4°   4°   _____

    d.   20°   6°   _____

In items 12–18, the statements are either true or false. If a statement is true, write in OK in the blank to the right. If it is false, underline the word(s) that makes the statement false, and write in the correct word(s) in the blank to the right.

12. The dew point depends upon the temperature of the air. _____

13. The dew point is measured in percent. _____

14. The dew point is reached when the air is cooled until it is saturated. _____

15. The dew point is reached when the relative humidity gets to 0%. _____

16. The dew point is usually higher when the air is dry. _____

17. The dew point changes only if the water vapor content of the atmosphere changes. _____

18. The dew point is important in the formation of clouds. _____

19. Define dew point. _____

_____

20. What does it mean when we say that the relative humidity is 100%?

_____

21. How can the relative humidity be increased without changing the amount of moisture in the atmosphere?

_____

22. Which of the three graphs in Figure 20-8 shows the correct relationship between temperature and evaporation?

_____

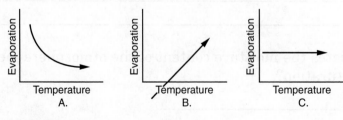

FIGURE 20-8.

☀ CHAPTER 20—LAB 2: A STATIONARY PSYCHROMETER

Introduction

Humidity can be expressed as a percent of saturation. This is known as the relative humidity. If the atmosphere contains as much moisture as it can at its present temperature, the relative humidity is 100%. If the relative humidity is 20%, at its present temperature the atmosphere could contain five times as much water as it now has.

The relative humidity is an important factor in determining the rate at which evaporation will occur. Water evaporates quickly when the relative humidity is low. When water evaporates it carries away 540 calories per gram of water that evaporates. That is a lot of energy. This transfer of latent heat energy causes the remaining water to become cooler. That is why you feel cold when you get out of the water on a dry, windy day. In evaporation, the most energetic molecules escape, leaving the cooler molecules behind.

On a day when the relative humidity is high you are more likely to feel "clammy." This is because sweat does not evaporate well under humid conditions. The phrase "the dog days of summer" describes hot, humid weather common in New York in summer months.

Objective

To take wet-bulb and dry-bulb thermometer readings, and use them to calculate the relative humidity.

Materials

Ring stand, 2 Celsius lab thermometers, 2 thermometer clamps, shallow dish with water, fan (such as a plastic lid), hollow section of shoe lace

Procedure

1. Secure the two thermometers to a ring stand as shown in Figure 20-9. One thermometer will hang freely so that it records the air temperature. This will be the dry bulb thermometer.

   The second thermometer is the wet-bulb thermometer. The bulb of this thermometer must be slipped inside a wet sleeve. (To keep the cloth wet, put one end into the dish of water as shown in Figure 20-9.) However, the bulb of the wet thermometer should hang above the water surface.

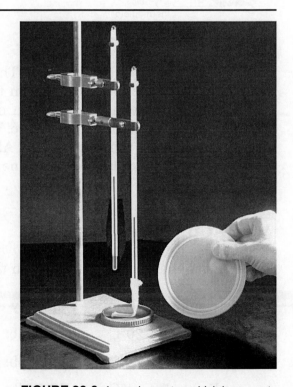

FIGURE 20-9. A psychrometer, which has a wet sock on one thermometer bulb, is used to determine humidity.

2. Let the thermometers sit for 3 minutes. Then record the temperature reading on each.

dry bulb _____ wet bulb _____

3. Now, fan the thermometers for 3 minutes. Record the two temperatures.

dry bulb _____ wet bulb _____

Find the difference between the wet-bulb and dry-bulb temperatures recorded in step 3.

_____

Next, open the *Earth Science Reference Tables* to the table at the bottom of page 12. To use this table, follow the vertical column down from the temperature difference until it meets the horizontal row of the air (dry-bulb) temperature. Where the column and row intersect, read the relative humidity in percent. (The most common mistake students make is to forget to calculate the wet-bulb depression!)

Meteorologists generally measure relative humidity with a hygrometer. For example, the sling psychrometer shown in Figure 20-10 is often used. Evaporation from the cover of the wet-bulb thermometer generally makes the wet-bulb temperature cooler than the dry-bulb (air) temperature.

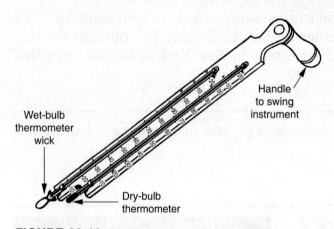

Handle to swing instrument

Wet-bulb thermometer wick

Dry-bulb thermometer

FIGURE 20-10. Sling psychrometer

To understand evaporation and condensation, you need to realize that scientists usually speak of net changes. That is, when scientists say water evaporates, they mean that there is more evaporation than condensation. Both changes always occur together. Therefore, when evaporation dominates, more moisture enters the atmosphere than moisture leaves the atmosphere. When condensation dominates, the greater change is moisture leaving the atmosphere. Although both changes occur simultaneously, scientists generally talk only about the one that dominates and ignore the smaller of the two changes.

Wrap-Up ▐▐▐▐▶

1. What is the relative humidity in the science room? _____

2. Determine the relative humidity under the conditions listed below:

    a. Dry bulb: 20°C   Wet Bulb: 15°C   Relative Humidity: _____

    b. Dry bulb: 24°C   Wet Bulb: 9°C   Relative Humidity: _____

    c. Dry bulb: −4°C   Wet Bulb: −6°C   Relative Humidity: _____

    d. Dry bulb: 25°C   Wet Bulb: 15°C   Relative Humidity: _____

3. Compared with the wet-bulb temperature before fanning, the wet-bulb temperature after it was fanned was _____.

   a. greater   *b.* the same   *c.* less

4. Why does fanning the wet-bulb thermometer make it cooler?

   _____

5. What does 100% relative humidity mean?

   _____

6. How much energy is needed to evaporate 1 gram of water?

   _____

7. As the humidity increases, what happens to the net rate of evaporation?

   _____

8. There is a puddle of water on a sidewalk. If there is no loss of water by evaporation, what is the relative humidity? _____

9. What weather instrument is used to determine the relative humidity?

   _____

10. What is the most common mistake that students make when using the Relative Humidity table?

    _____

11. What substance is the primary reservoir of heat energy in Earth's atmosphere?

    _____

12. What health problems are caused by very low humidity, and what problems by very high humidity?

    _____

    _____

My Notes

CHAPTER 20—LAB 3: HOMEMADE CLOUDS

Introduction

Clouds are large accumulations of tiny ice crystals or water droplets so small they remain suspended in the air for a long time. If you ever wonder what it is like inside a cloud, just look around you when there is a fog. Clouds form when the air temperature falls below the dew point and the air contains tiny condensation nuclei (dust, smoke, etc.) for water to condense onto. Most clouds are high enough that they form by deposition as tiny ice crystals. If they grow large enough to fall into warmer air, they may become rain.

Objective

To observe cloud formation and evaporation and investigate factors that influence these changes.

Materials

A large glass jar or large beaker, hot and cold tap water, 1-gallon plastic bag or a shallow metal pan, 100–200 mL of ice or snow, match

Procedure

1. Run about 2 cm of hot tap water into the bottom of a large glass jar or beaker. Swirl the jar to reduce condensation inside.

2. Place a few ice cubes or a handful of snow in the plastic bag or a shallow pan. Add roughly 100 mL of cold tap water.

3. Place the ice bag or pan over the jar as shown in Figure 20-11. Can you see a cloud forming inside the jar?

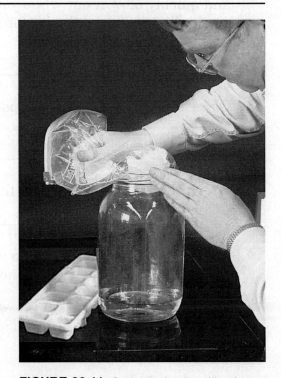

FIGURE 20-11. Cover the hot jar with a bag of ice water.

FIGURE 20-12. Extinguish the match and let smoke into the hot jar.

FIGURE 20-13. What happens to the cloud when it escapes?

4. Remove the bag of ice water from the jar. (Save the bag of ice water.) Light a match and quickly blow it out while you hold it inside the jar. (See Figure 20-12.) Hold the smoldering match in the jar for a few seconds. What do you see mixing with the air inside the jar?

_____

5. Cover the jar with the bag or pan of ice water as you did before. What happened this time?

_____

6. Watch the air circulate inside the jar. What kind of heat flow are you observing?

_____

7. Remove the bag of ice water to watch the cloud escape and disappear as in Figure 20-13. Where has the cloud gone?

_____

Wrap-Up ▶▶▶

1. What two things were supplied by the hot water in the jar?

_____ and _____

2. Why was smoke needed?

_____

3. What was the purpose of the bag of ice water?

_____

4. To what temperature did the air inside the jar need to be cooled to make a cloud?

_____

5. If water vapor is invisible, how can we see clouds?

_____

6. What change in state causes the formation of a cloud made of water droplets?

_____

7. Most clouds are made of ice. What process forms them? _____

8. Do the two changes in state mentioned in items 6 and 7 use or release heat energy?

_____

9. According to the *Earth Science Reference Tables,* how much latent energy is released when a liquid cloud with a mass of 1 gram is formed?

_____

10. Heat energy involved in a change in state of water, such as freezing or boiling, can be calculated with the following formula:

Heat energy gained or lost = mass (in grams) × *heat of vaporization or heat of fusion*

Calculate the amount of energy required to evaporate 5 grams of water. Show your work below. (You must show the algebraic formula, substitutions, numerical result, and units.)

11. Calculate the energy released when 20 grams of water freezes.

12. What is Earth's source of energy used to evaporate approximately 16 million tons of water every second?

_____

13. What changes in state require the addition of heat energy?

_____

14. What two changes in state release energy?

_____

15. Deposition changes water vapor directly into ice. In fact, this is how most clouds form. How many calories are released for each gram of water that is deposited as ice crystals?

_____

_____

_____

CHAPTER 20—LAB 4: THE HEIGHT OF CLOUDS

Introduction

Have you ever wondered what it would be like if you were inside a cloud? Perhaps you have been. You may have been in an airplane, or you may have been outside on a foggy day. When a cloud forms at ground level, we call it fog.

Most clouds form when water vapor changes directly to ice crystals high in the atmosphere. This process is called deposition. (Only the lowest clouds are composed of tiny water droplets.) Deposition or condensation can occur only if tiny particles called condensation nuclei are suspended within the atmosphere. Those particles provide surfaces on which deposition and condensation can occur.

You may have noticed that the bottoms of the clouds are often relatively flat as shown in Figure 20-14. A cloud forms when air rises to a height at which the air is cooled below the dew point. As air rises, it expands. A temperature change in a gas caused by expansion or compression is known as an adiabatic (a-DEE-e-BAT-tic) temperature change. Expanding air pushes outward, cooling because this releases the energy used to compressed it. If air cools below the dew point, a cloud may form. The base of the cloud forms at the height at which air is cooled to the dew point.

FIGURE 20-14. Fair weather cumulous clouds over the land.

Warm, moist air rises into the atmosphere because it is relatively low in density. However, when air rises, expansion causes cooling. If moist air cools below the dew point, and small particles from dust, smoke, or saltwater spray are present within the atmosphere, deposition or condensation will form a cloud. Cloud droplets grow by joining with other droplets. This is called coalescence. When the drops become large enough, they fall out of clouds to the ground. As raindrops fall, they continue to merge with other drops, growing steadily larger.

If you know how much a parcel of air must be cooled to reach the dew point, you can use a chart to determine how high air must rise for a cloud to form.

Objective

To determine the dew point of the air in your classroom.

Materials

Shiny silver (metal) cup, thermometer, plastic container of snow or crushed ice

Procedure

1. Measure and record the air temperature in the classroom. _____ °C

2. Fill the metal cup half way with room temperature water.

3. Obtain several ice cubes in a plastic container.

4. Do this step very slowly. Stir with the thermometer and constantly feel the outside of the cup for condensation. As shown in Figure 20-15, add snow or ice *a little at a time* until you can see or feel dampness (dew) on the outside of the cup. Dampness will indicate that the cup's surface is at or below the dew point. Work slowly. The temperature at which dew (water) first starts to form on the outside of the cup is

   _____ °C

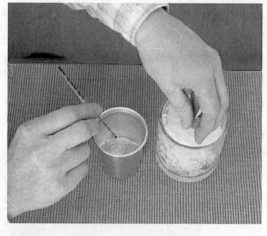

FIGURE 20-15. Add the crushed ice very slowly and feel carefully for wetness on the can.

5. Now that you know what to look for, try it again for more accuracy.

   Second determination of dew point is

   _____ °C

6. On Figure 20-16, write a small T along the bottom of the graph to indicate the air temperature found in step 1.

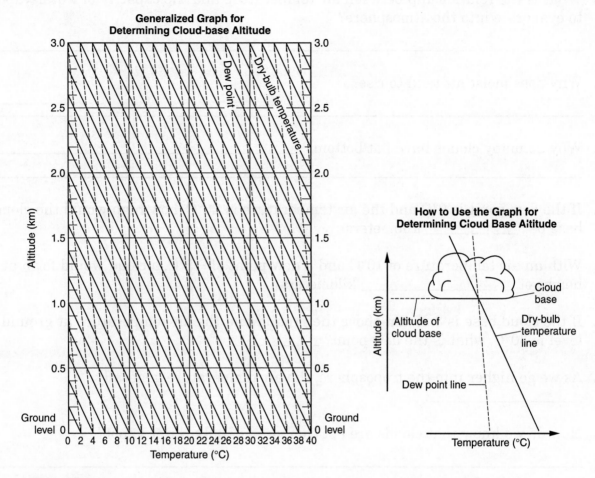

FIGURE 20-16.

7. Next, write a D to indicate the dew point, also the bottom of this graph.

8. Follow the trend of the solid lines up from the T until it crosses the trend of the dotted lines above the D. That intersection shows the height of the clouds. Read the height on the scale at the right side of the chart. Today's cloud height would be

_____ km.

9. You can use the same apparatus to observe a model of another change high within the atmosphere. This time, half fill a dry silver cup with cold water. Then add ice until the cup is nearly full. What happens, as the drops of water on the outside of the cup grow very large?

_____

10. What two processes within clouds are being modeled here?

_____ and _____

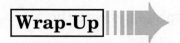

Wrap-Up

1. What is the relationship between air temperature and the capacity of water vapor to evaporate into the atmosphere?

   _____

2. Why does moist air tend to rise?

   _____

3. Why do many clouds have flat bottoms _____

   _____

4. If the dew point is 0°C and the air temperature is 20°C, how high would the cloud base be? _____ kilometers

5. With an air temperature of 10°C and a dew point of −5°C, a cloud would form at a height of _____ kilometers.

6. If the cloud base is 3.25 km above the ground and the air temperature at ground level is 30°C, what is the dew point? _____

7. As we go higher into the troposphere, the air temperature generally _____.

8. How do we know that clouds are not composed of only water vapor?

   _____

9. What is the source of the water that formed on the outside of the cold cup?

   _____

10. What word describes the cooling of a gas caused by expansion? _____

11. How can air be warmed without adding heat energy? _____

12. Most clouds form by deposition. What do we mean by deposition?

    _____

13. If high summer clouds are composed of ice crystals, why does it rain rather than snow?

    _____

14. What is a cloud?

    _____

CHAPTER 20—LAB 5: INQUIRY INTO THE RELATION BETWEEN HEAT AND TEMPERATURE

Introduction

In this lab, you are to design a procedure and decide on what materials you will need to complete the objective.

CAUTION: Use appropriate safety equipment and wear protective clothing. Both solid objects and hot water can cause serious burns. Work safely and conduct yourself in a cautious manner. If you need to move hot materials, your teacher can help you.

Objective

Determine the answer to the following question. If you heat a small amount of water and a larger amount of water, initially at the same temperature, for the same time, will their final temperature be the same? Explain your response.

Materials

Present your list of materials to your teacher for approval.

Procedure

Devise a procedure and submit it to your teacher for approval.

Write a lab report in which you give your objective, materials, procedure, and conclusions.

My Notes

Chapter 21
Air Pressure and Winds

CHAPTER 21—SKILL SHEET 1: AIR PRESSURE AND WEATHER

There are many stories about farmers looking up at the sky and predicting the weather more accurately than today's professional meteorologists. This brings up two interesting questions. First, how much do all the modern devices weathermen use really improve their forecasts? Second, can simple observations lead to good weather forecasts? The procedure that follows should help you answer the second question.

If you know the wind direction, and how the atmospheric pressure is changing, does this help you predict future weather? You will investigate this question by constructing a unique type of graph of weather data.

TABLE 21-1. Weather Data Table

Date	Pressure	Wind	Weather	Date	Pressure	Wind	Weather
4/01	High +	310°	S & C	4/13	Med	220°	S & W
4/02	High	260°	S & C	4/14	High −	225°	S & W
4/03	Med	245°	S & W	4/15	High −	235°	S & W
4/04	Low +	195°	S & W	4/16	High	340°	S & C
4/05	Low	160°	R & W	4/17	High −	325°	C & C
4/06	Med	250°	C & W	4/18	High	300°	S & C
4/07	High	320°	S & C	4/19	Med	20°	C & C
4/08	Med	270°	C & W	4/20	Low +	110°	R & C
4/09	Low +	40°	R & C	4/21	Low	190°	C & W
4/10	Low	75°	R & C	4/22	Low −	200°	R & W
4/11	Low	130°	C & W	4/23	Low +	230°	S & W
4/12	Med	210°	C & W	4/24	Med	260°	S & C

In the data table, wind direction is given in degrees that correspond to the circle graph included in this skill sheet. The following abbreviations are used R & C = rainy and cool,

C & C = cloudy and cool, S & C = sunny and cool, R & W = rainy and warm, C & W = cloudy and warm, and S & W = sunny and warm. Use the data above and the weather symbols in the key on the circular graph to plot the proper symbol at a position determined by the wind direction and air pressure data. (The calendar dates should not be shown on the graph.)

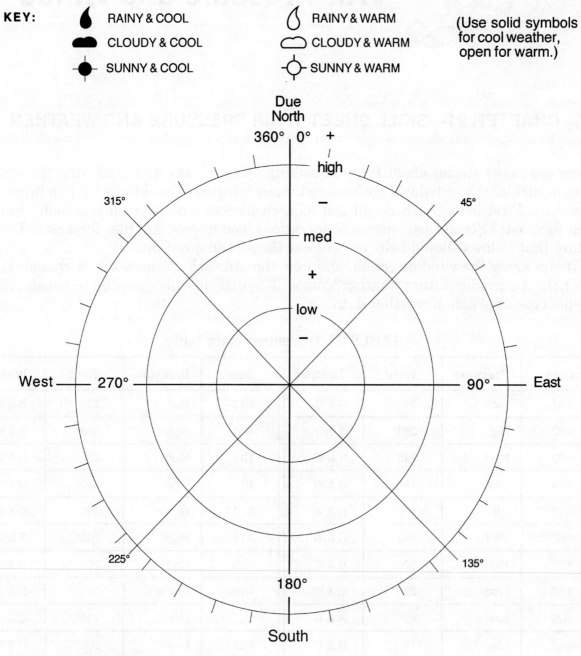

KEY:

     RAINY & COOL   RAINY & WARM

     CLOUDY & COOL   CLOUDY & WARM

     SUNNY & COOL   SUNNY & WARM

(Use solid symbols for cool weather, open for warm.)

FIGURE 21-1. Polar graph paper.

Wrap-Up

1. ***a.*** What is the azimuth (direction) angle of east? _____

   b. What is the azimuth (direction) angle of southwest? _____

2. Winds are labeled by the direction the wind comes from. Based on your graph, what appears to be the most common (prevailing) wind direction?

   _____

3. What kind of weather usually occurs when the wind is blowing from the northeast?

   _____

4. What wind direction usually brings low pressure? _____

5. Why is moist air associated with low air pressure?

   _____

6. What two characteristics of the weather are associated with high atmospheric pressure?

   _____ and _____

7. As water evaporates into the atmosphere, what usually happens to the density of the atmosphere? Why?

   _____

   _____

8. Of these 24 days, what percent of the time did the winds come from the northeast?

   _____

9. What fraction of the time did the winds blow from the southwest? _____

10. This procedure is similar to the methods used by the news media to measure the popularity of political figures. In what way is it similar?

    _____

11. This kind of graph paper is called "polar" graph paper. Why?

    _____

12. Suggest another scholarly use for this type of graph paper grid.

    _____

    _____

My Notes

CHAPTER 21—SKILL SHEET 2: LAND AND SEA BREEZES

Winds blow from regions of higher pressure toward regions of lower pressure. At sea level above each square inch of Earth's surface is a column of air that weighs nearly 15 pounds. That column of air is the cause of barometric air pressure. However, Earth's rotation and the Coriolis effect also influence the wind direction.

Why is the shore so popular in hot summer weather? During the day, when the sun shines, the land heats up more than water. With its high heat capacity, water is slow to heat or cool. You may recall that when air is heated, it expands and becomes less dense. When warm, light air over the land rises, the cooler air from the ocean blows in to replace it; the result is a local wind. (See Figure 21-2.)

Not only does this temperature differential crate a pleasant breeze, the wind is relatively cool. Continued heating of the land sustains these winds until the sun sets and the land cools. Then the breeze reverses. (See Figure 22-3.)

The night breeze blows from the land back out to sea. This is because at night, the land cools rapidly, but the water holds its warmth. Now the ocean is warmer than the land, and the warm ocean air rises, causing the air from over the land to move out to sea. Heat flow caused by differences in the density of a fluid is known as convection.

These localized winds make the shore a popular place in hot summer weather. As long as there are no strong regional weather systems to overpower these land and sea breezes, these localized winds moderate the summer temperatures at the shore.

FIGURE 21-2. A sea breeze blows onto the land.

FIGURE 21-3. A land breeze blows out onto the water.

1. Winds always blow from _____ pressure toward _____ pressure.

2. What instrument is used to measure atmospheric pressure? _____

3. What causes low pressure over the land during the day? _____

4. What are the U.S. and metric units of barometric pressure? (The *Reference Tables* may help you.) U.S. _____, metric _____

5. What term is applied to the circulation of energy by density currents in a fluid?

   _____

6. What is a sea breeze? _____

7. On Figures 21-2 and 21-3, print an *H* to show where the surface atmospheric pressure is relatively high and an *L* to show there the surface pressure is relatively low.

8. Why does the ocean temperature change less than the land temperature?

   _____

9. According to the *Earth Science Reference Tables* (page 14):

   29.5 inches of mercury = _____ mb, 998 mb = _____ inches of mercury

   30.21 inches of mercury = _____ mb, 1016 mb = _____ inches of mercury

10. What is the weather usually like when atmospheric pressure is high?

    _____

11. Why is wind, like force, considered a vector quantity?

    _____

Chapter 22
Weather Maps

Chapter 22—Skill Sheet 1: Fronts

Air masses are bodies of air that are relatively uniform in temperature and humidity. These bodies of air take on the characteristic of their geographic place of origin. For example, in summer, an air mass moving into New York from the South Atlantic Ocean is likely to be moist and warm. This would be a maritime tropical air mass. Most air masses are classified according to their humidity as maritime (moist) or continental (dry). The temperature is indicated by tropical, polar, or arctic. Arctic air masses move into in New York State only in the winter. They bring our coldest winter weather.

Figure 22-1 is a satellite photograph of a low-pressure system. Low-pressure systems, also known as cyclones, are zones of air convergence. At the center of a cyclone, relatively warm, moist air rises. It rises because warm air and moist air are lower in density than cool air and dry air. As the lighter air at the center of the cyclone rises, different air masses move into (converge on) the low-pressure center.

FIGURE 22-1. This counterclockwise and inward circulation of winds is a characteristic of all Northern Hemisphere mid-latitude cyclones.

High-pressure systems, also known as anticyclones, are zones of air divergence. They form where descending cool and/or dry air spreads out at Earth's surface like pancake batter poured on a griddle. The boundaries (interfaces) between different bodies of air are known as fronts. Most weather changes are associated with the passage of fronts.

1. What are the temperature and humidity characteristics of a maritime polar air mass?

_____

2. What is the most likely source region for a continental polar air mass blowing into New York State?

_____

3. How is a continental tropical air mass abbreviated in the *Earth Science Reference Tables*?

_____

4. What four-word name for a cyclonic weather system tells you that air masses are drawn in toward the center?

_____

5. What air motion causes cloud formation, and causes the winds to converge into cyclones?

_____

6. Define weather front. _____

7. When water vapor is added to the atmosphere, how does this affect the air's density?

_____

There are four types of frontal boundaries. Each brings its own characteristic weather. The intensity of these fronts is determined by the difference in atmospheric pressures within the cyclone. The atmospheric pressure depends on the temperature and humidity of the air masses. Figure 22-2 illustrates the development of a cyclone.

Warm fronts are usually preceded over a period of several days by thickening clouds. Wispy cirrus clouds may move lower and change to layered stratus clouds. As the front moves closer, steady rain and fog are common. The passage of the front is marked by a change

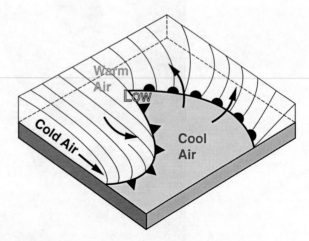

FIGURE 22-2. This developing cyclone has an advancing cold front, shown by the triangles, and an advancing warm front, shown by the half circles.

to warmer but usually hazy weather. Warm fronts often bring maritime, tropical air masses in the Northeast.

Cold fronts generally bring cool, crisp, and dry weather. They pass quickly, sometimes within an hour or two. A few, puffy cumulus clouds rapidly build into massive clouds of great height as a cold front approaches. Precipitation can be intense, but brief. Summer cold fronts often bring thunderstorms. As the front passes, the temperature drops, while air pressure suddenly increases, because the cool dry air following the cold front is relatively dense.

Although fronts may pass quickly, they are not the sharp interfaces you might conclude from looking at a weather map. Most fronts are zones in which temperature and other atmospheric variables change rapidly. The severity of frontal weather depends on the difference in temperature and humidity between the air masses. Thunderstorms and tornadoes are often associated with cold fronts. However, a weak front may just cause cloudy weather without precipitation.

8. Complete the table below based upon the text you just read.

	Warm Front	**Cold Front**
Relative speed		
Temperature change		
Precipitation		
Barometric pressure		
Air mass that follows a strong front		

9. Why do warm fronts advance more slowly than cold fronts?

_____

10. On what does the severity of frontal weather depend?

_____

Some fronts do not move. These fronts are called stationary fronts. There is very little mixing of air masses along a stationary front. The polar front that separates the polar zone of easterly winds from the mid-latitude westerlies is often a stationary front. Winds generally blow in opposite directions along a stationary front. Nevertheless, as the name implies, the frontal boundary of a stationary front does not move.

An occluded front develops when a fast-moving cold front overtakes a warm front. When an occluded front develops, cold air pushes the warm air aloft. As shown in Figure 22-3, the denser, drier, cooler air pushes up the warm air. The line with the alternating half circles and triangles is the occluded front. Of these four weather fronts, this is the only one in which the frontal boundary does not touch the ground. The weather near an occluded front is usually cloudy or rainy for a relatively long period.

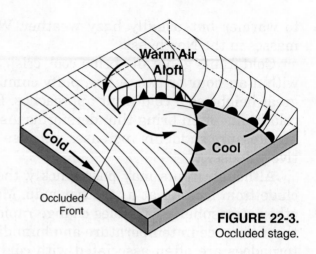

FIGURE 22-3.
Occluded stage.

11. When warm air is pushed up by cooler air closing in from both sides, what kind of front results?

_____

12. Name the front that generally stays in the same place. _____

13. What front separates the cold polar easterly winds from the mid-latitudes westerlies?

_____

14. How does an occluded front form? _____

Figure 22-4 shows how an eddy (swirl) develops in the polar front. The eddy grows into a mid-latitude cyclone as a zone of air convergence forms and draws in different air masses. The boundaries between these air masses are weather fronts.

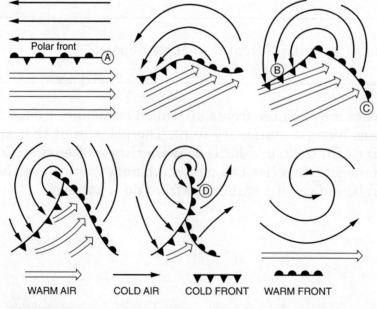

FIGURE 22-4. Mid-latitude cyclones start and end as an eddy in the polar front.

What happens to the air that is drawn into a low-pressure system? It rises and spills out in the upper atmosphere. Therefore a zone of convergence at the surface sits below a zone of divergence (a high) in the upper atmosphere. Conversely, a high-pressure area at the surface sits below a low-pressure center high in the atmosphere.

15. Write the type of front shown by each letter A through D in Figure 22-4.

A. _____

B. _____

C. _____

D. _____

Figure 22-5 illustrates a mature cyclone. The first diagram is a map view of the low-pressure system (cyclone). The three cross-sectional views show the three types of fronts found in a mature cyclone.

16. Name the three kinds of fronts in Figure 22-5.

E. _____

F. _____

G. _____

Figure 22-6 shows the generalized circulation pattern of Earth. Notice that the polar front has been labeled.

In winter, the polar front moves south, bringing the cold, crisp conditions of winter weather. Winter thaws occur when the polar front temporarily moves back north. In the summer, the polar front moves well to the north, and it is seldom a factor in our summer weather.

17. Why do occluded fronts bring a relatively small change in temperature?

_____

_____

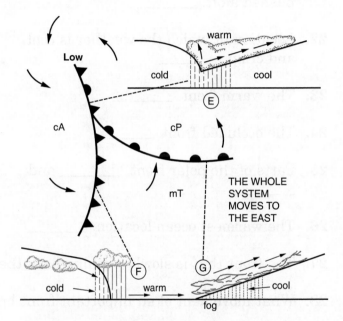

FIGURE 22-5. A map view of a mature cyclone with vertical cross sections through the fronts.

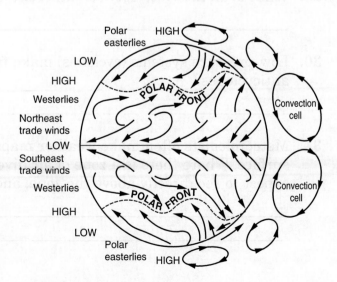

FIGURE 22-6. Global wind belts.

Answer the following by matching each item with the correct letter from the map in Figure 22-7. Note that three letters label frontal lines and the other three do not.

18. The cold front. _____

19. The center of the warm air mass. _____

20. Location at which it will soon to get much warmer. _____

21. Front with the whole warm air mass pushed aloft. _____

22. Location at which the weather is cool and crisp. _____

23. The warm front. _____

24. The occluded front. _____

25. Parts of the polar front. _____ and _____

26. The warmest ocean location. _____

27. The front that is slowly pushing back the cold air. _____

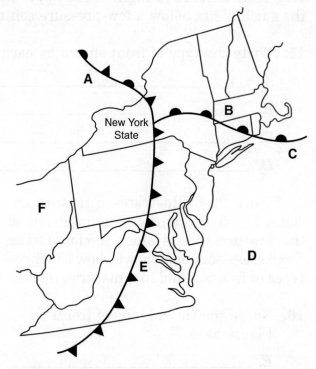

FIGURE 22-7. A weather system in the Middle Atlantic states.

28. What movement of an important front brings us our cold winter weather?

_____

29. What is an interface between different air masses called?

_____

30. Low-pressure systems (cyclones) make fronts. Why do fronts not form in anticyclones?

_____

31. Meteorologists often make weather maps at different elevations above Earth's surface. Where there is a zone of air divergence at Earth's surface, what would you expect to see directly above the zone, and why?

_____

_____

CHAPTER 22—LAB 1: JANUARY 2004

Introduction

Meteorologists know that the upper atmosphere jet stream plays an important role in forming and guiding the path of mid-latitude weather systems (cyclones). When the jet stream, which usually blows from the west, wanders north or south, unusually warm or cold weather may result. The weather of January 2004 was a good example.

Materials

Figures 22-8 through 22-10, pencil

Objective

In this activity you will be constructing two isotherm maps at an interval of 10°F.

Procedure

The first isotherm map will show the normal January conditions. The arrows on both figures indicate the directions of the prevailing (most common) winds for this month.

1. On Figure 22-8 draw the normal January isotherms.

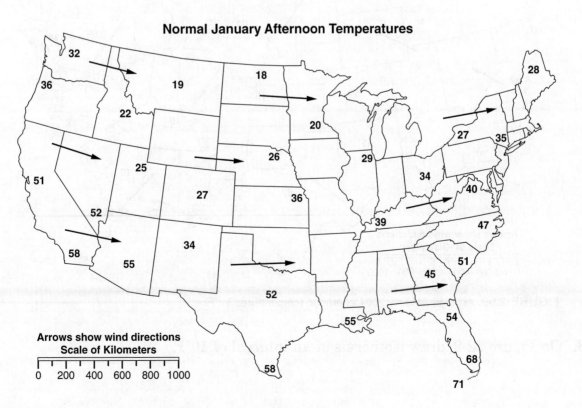

FIGURE 22-8. Normal January afternoon temperatures in °F.

2. What do we call lines connecting points of the same temperature?

_____

3. According to Figure 22-8, what is the most common wind direction in the United States? (Remember that winds are labeled by the direction the wind blows *from.*)

_____

4. As latitude increases, what usually happens to the temperature?

_____

5. Why do the isotherms dip southward in the western part of the United States?

(*Hint:* Think of the geographic features that cause the winters to be relatively cold in much of this part of the country.)

_____

The second isotherm map, Figure 22-9, illustrates the unusual temperature conditions that occurred in January of 2004.

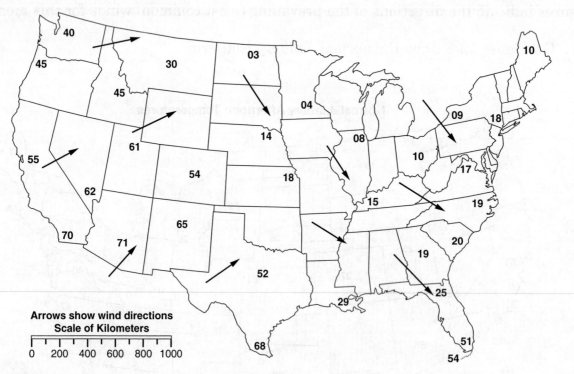

FIGURE 22-9. Average January 2004 afternoon temperatures in °F.

6. On Figure 22-9, draw isotherms at an interval of 10°F.

LABORATORY MANUAL

7. When you have finished drawing the isotherms, ask your teacher to help you draw the warm front and the cold front on Figure 22-9. Use the symbols shown in Figure 22-10.

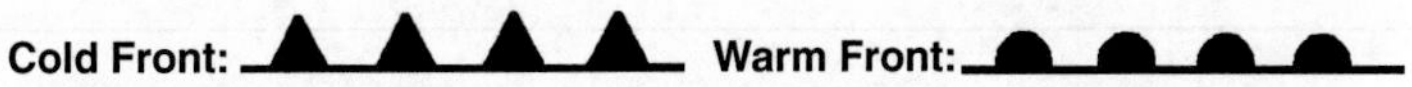

Cold Front: ▲▲▲   Warm Front: ●●●

FIGURE 22-10. Warm and cold front symbols.

8. To the nearest whole degree, what are the latitude and longitude angles of New York City?

_____ and _____.

9. How was January 2004 different from normal winter weather in the eastern part of the United States?

_____

10. How was this month's weather unusual in the western states?

_____

11. Why do you think that the winter of 2004 was so unusual? (*Hint:* Read the introduction to this lab.)

_____

12. As shown in Figure 22-9, what kind of air mass was moving into Florida?

_____

13. In 2004, what kind of air mass is located in the southwestern United States? _____

    a. cT   *b.* cP   *c.* mT   *d.* mP

14. Why would you end your isolines near the shorelines of the oceans?

_____

15. What was the average January 2004 afternoon temperature at Albany, New York?

_____

16. The number that appears in southeastern New York State represents the temperature in New York City. In degrees Celsius, what was the average January 2004 afternoon temperature in New York City?

_____

17. Calculate the 2004 temperature gradient from Key West, off the southern tip of Florida, to Buffalo, New York.

_____

_____

_____

CHAPTER 22—SKILL SHEET 2: WEATHER STATION DATA

Use the *Earth Science Reference Tables* (page 13) to help you decode the weather symbol below.

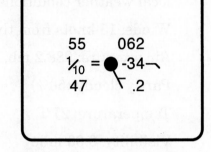

FIGURE 22-11. Station model with data.

1. Present weather: _____

2. Air temperature: _____. Dew-point temperature: _____

3. Wind direction: (from the . . .)

   _____

4. Wind speed: _____

5. Atmospheric pressure: _____

   and (rising or falling how much): _____

6. Cloud cover: _____

Your teacher will supply you with a map from the United States Weather Bureau. Choose one city on the map that is relatively easy to read.

7. What is the date of the map that you have selected? _____

8. Select one city for which the weather data is readable.

   City: _____, State _____

9. What is the:

   a. Wind direction: _____

   b. Wind speed: _____

   c. Atmospheric pressure: _____

   d. Percentage of cloud cover: _____

   e. Present temperature: _____

   f. Dew point: _____

10. Most weather systems move across the United States from _____ to _____.

11. Based on the generalized direction of movement, predict the weather for the next several days at the city you have selected.

    _____

12. Based on this map, predict the weather for the next several days at Albany, NY.

_____

13. Use the box in Figure 22-12 to draw a weather symbol to represent the following local weather conditions:

Winds: 15 knots from the northeast

Air pressure: 988.2 mb, and up 2.5 mb in the past 3 hours

Partly cloudy (50%)   Snow flurries

Temperature: 27°F   Dew point: 24°F

Visibility: 0.33 mile   Precipitation: $\frac{1}{20}$ inch
                           (water equivalent)

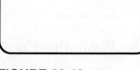

FIGURE 22-12. Station model for question 13.

Note the pattern of fronts around the low-pressure centers in Figure 22-13. You are very likely to see similar patterns on other weather maps.

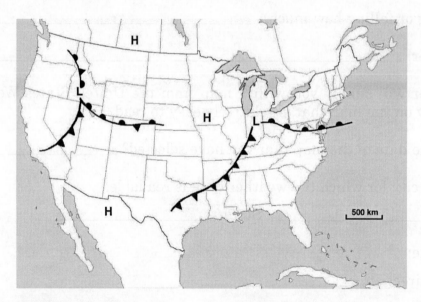

FIGURE 22-13. Sample pattern of fronts in the contiguous United States.

14. On a weather data map, how can you tell where the temperature and air pressure measurements were taken?

_____

15. Make a weather map station model to show surface conditions on an imaginary planet. Share the station model with a classmate, and have him or her describe the weather conditions.

CHAPTER 22—LAB 2: SYNOPTIC WEATHER MAPS

Introduction

Have you ever read a synopsis of a book? A synopsis is a brief summary of the story. Similarly, a synoptic weather map is a synopsis (summary) of the weather. It shows a variety of field quantities on a single map. Figure 22-14 is an example of a synoptic weather map. The synoptic weather map is one of the most important tools that a meteorologist uses for making weather forecasts. With a current map and knowledge of past weather changes, a meteorologist can often make good predictions for the next few days.

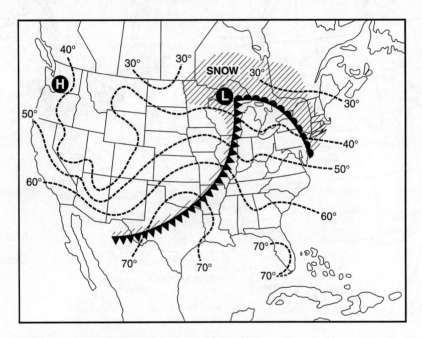

FIGURE 22-14. USA weather map.

You should know that a field is a region in which one or more variables can be measured at different places. In this lab, the maps that you will be drawing will show the following five variables recorded in the region shown on the map: barometric pressure, temperature, precipitation, wind direction, and cloud cover.

Objective

Draw synoptic weather maps for a given set of conditions.

Materials

USA weather data maps: Figure 22-15: **Map A**, and Figure 22-17: **Map B**, pencil, transparent highlighter, colored pencils

Procedure

This lab must be done neatly. Use a pencil to draw your isolines. Use a transparent highlighter pen or colored pencils to shade in the areas of clouds and precipitation. Your lab will be graded on *accuracy, readability, and neatness.*

Map A

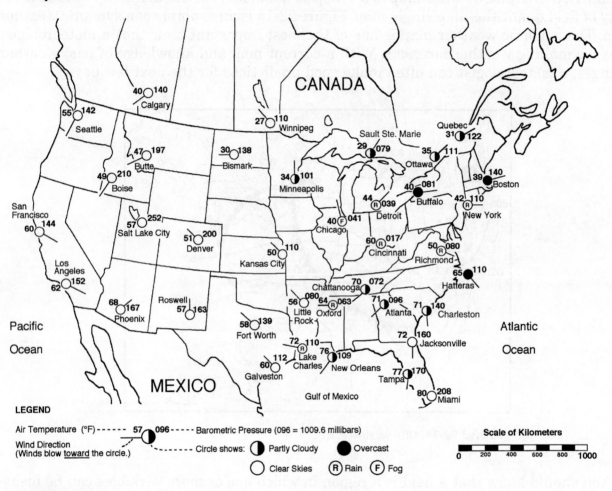

FIGURE 22-15.

1. On Map A, draw isobars (lines of equal air pressure) at an interval of 4 mb. Start with 040, then 080, etc., until you have covered all the air pressure data.

2. Label the center of the cyclone *L* and the center of the anticyclone *H*. (A cyclone is a low-pressure system and an anticyclone is a high-pressure system.)

3. Show the wind direction at each station by extending the arrows at each station as shown here. Be sure that you darken each line and construct the points of the arrows.

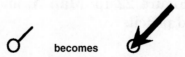

FIGURE 22-16. How to draw arrows from wind direction sticks.

Map B

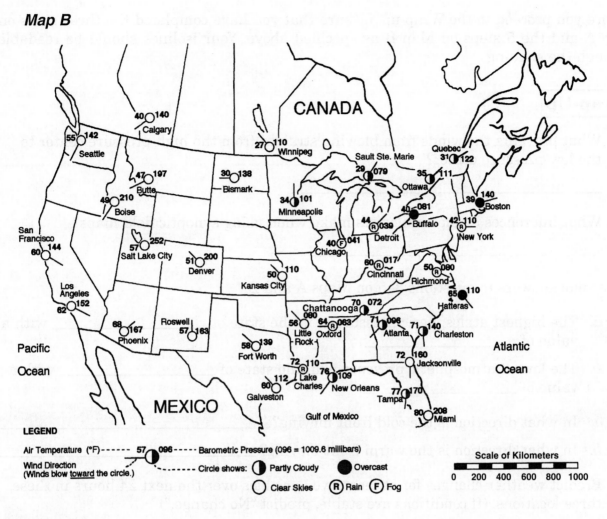

FIGURE 22-17.

4. On Map B, write in the H and the L in the same places where they are on Map A.

5. Draw isotherms at an interval of 5°F. Start with 30°.

6. Lightly outline the one region that has any cloud cover at all. That is, draw a light line to separate the areas with partial or complete cloudiness from the clear areas. Then, use a colored pencil or a highlighter to lightly shade this cloudy area. (Be sure to include any areas with rain or fog in the cloudy region.)

7. Outline the one area with rain or fog. Shade that area a little darker, or with a different color.

   (The numbers and isolines should be visible and readable through the shading.)

8. Draw the warm front (Figure 22-18A) and the cold front (Figure 22-18B) along the places where the isotherms are close together. Look at the wind direction to decide how and where to draw the fronts. The circles or triangles should point in the direction that the front is moving. (*Hint:* Both fronts extend out of the low-pressure center, as on the weather map in Figure 22-14.)

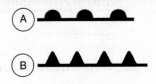

FIGURE 22-18.

Before you proceed to the Wrap-up, be sure that you have completed the three steps on Map A and the 5 steps on Map B as specified above. Your isolines should be readable through the shading.

Wrap-Up ▏▏▏▶

1. What prevents the winds from blowing straight from the high-pressure center to the low-pressure center?

   _____

2. What inferences do meteorologists make when using synoptic field maps?

   _____

Base your answers to questions 3–5 on Maps A and B.

3. **a.** The highest atmospheric pressure is in the state of _____ with a value of _____ .

   b. The lowest atmospheric pressure is in the state of _____ with a value of _____ .

4. **a.** In what direction is the cold front moving? _____

   b. In what direction is the warm front moving? _____

5. Predict weather changes for at least two variables over the next 24 hours in these three locations. (If conditions are stable, predict "No change.")

   a. Chattanooga, TN:

   _____

   _____

   b. New York, NY:

   _____

   _____

   c. Denver, CO:

   _____

6. Attach a copy of a current US weather map from a newspaper or another source and write below it the current conditions in Phoenix, Arizona.

   _____

7. What field quantities are shown on the weather map you selected for question 6?

   _____

   _____

LABORATORY MANUAL

Chapter 23
Weather Hazards and the Changing Atmosphere

CHAPTER 23—LAB 1: USING HURRICANE TRACKING DATA

Introduction

Hurricanes are tropical storms with sustained winds of 74 miles per hour (120 km/hr or 64 knots.) These winds are often accompanied by torrential rain, tornadoes, and coastal flooding.

Hurricanes get their destructive energy through condensation (cloud formation). For each gram of water vapor that changes into cloud droplets, 540 calories of energy are released. Accelerating cloud formation causes the storm to strengthen, changing from a tropical depression, to a tropical storm, and finally a hurricane. As hurricanes move across the ocean, they often grow stronger. However, they lose energy when they move over cooler waters at higher latitudes, or over land. The latent energy of moist, tropical, oceanic air maintains the strength of hurricanes. Hurricanes also lose energy to friction with vegetation and the land surface.

FIGURE 23-1. Hurricane Isabel approaches the East Coast in 2003.

Materials

Hurricane data, map, pencil

Objective

In this lab, you will use data collected by meteorologists to plot the track of a hurricane on a map.

Procedure

1. Use the data on page 263 or data supplied by your teacher to plot the path of a hurricane on the map in Figure 23-2. You may save time by plotting only the first position of each day. Label midnight (00) of each new day with the date. (For example, the first midnight, August 15, could be labeled 08/15.)

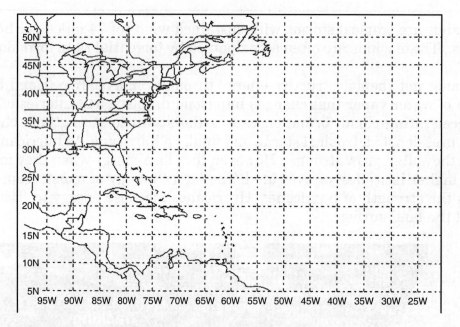

FIGURE 23-2. Atlantic hurricane tracking map.

2. Draw parentheses around the area where this tropical storm was a hurricane (category 1 or above) and label this line "Hurricane." Then make an X where the storm reached maximum strength. Label the X "Category N," where N = the maximum strength of this hurricane.

Hurricane Tracking Data

The data table that follows is a list of the position of the center of circulation (the eye) of tropical cyclone Andrew from August 14 to 27, 1992. The time (Hr) is UTC, that is, 18Z = 12.00 P.M. EDT. The latitudes listed are North of the Equator, and longitudes are West of the Prime Meridian. Wind speeds (WndSpd) are in knots (1 knot = 1.15 mph, 120 kt =

138 mph). An estimate of the category based on the Saffir/Simpson Scale is listed when the winds exceed 64 knots (74 mph). (Data from H. Friedman, Hurricane Research Division, Atlantic Oceanographic and Meteorological Laboratory, NOAA.)

TABLE 23-1. NOAA Tracking Data for Tropical Cyclone Andrew, August 14–24, 1992

Day	Hr	Latitude (°N)	Longitude (°W)	WndSpd (knots)	Day	Hr	Latitude (°N)	Longitude (°W)	WndSpd (knots)	Category
14	18	9.7	21.0	25	21	06	23.8	63.3	35	
15	00	9.7	21.8	25	21	12	24.4	64.1	40	
15	06	9.2	23.7	25	21	18	24.8	65.0	45	
15	12	9.4	26.2	25	22	00	25.2	65.9	50	
15	18	9.6	28.3	25	22	06	25.6	67.0	55	
16	00	9.8	30.0	25	22	12	25.8	68.3	65	1
16	06	10.1	31.6	30	22	18	25.7	69.7	75	1
16	12	10.5	33.3	30	23	00	25.6	71.1	95	2
16	18	10.9	35.3	30	23	06	25.5	72.5	105	3
17	00	11.2	37.4	30	23	12	25.4	74.1	110	3
17	06	11.7	39.6	30	23	18	25.4	75.7	130	4
17	12	12.3	41.9	35	24	00	25.4	77.5	120	4
17	18	13.0	44.2	40	24	06	25.5	79.5	130	4
18	00	13.6	46.2	45	24	12	25.5	81.4	120	4
18	06	14.1	48.0	45	24	18	25.8	83.2	120	4
18	12	14.6	49.9	45	25	00	26.2	85.0	120	4
18	18	15.4	51.8	45	25	06	26.6	86.7	100	3
19	00	16.3	53.5	45	25	12	27.2	88.2	100	3
19	06	17.1	55.3	45	25	18	27.8	89.5	105	3
19	12	17.9	56.9	40	26	00	28.5	90.5	100	3
19	18	18.8	58.4	40	26	06	29.2	91.2	85	2
20	00	19.8	59.5	40	26	12	30.0	91.7	60	1
20	06	20.8	60.2	40	26	18	30.8	92.2	40	
20	12	21.7	66.7	35	27	00	31.5	91.8	30	
20	18	22.5	61.5	35	27	06	32.1	90.6	30	
21	00	23.2	62.5	35	27	12	32.9	89.8	30	

Wrap-Up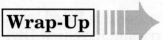

1. What must occur for a tropical storm to become a hurricane?

2. Where (latitude and longitude) was this storm when it became classified as a hurricane?

3. This hurricane was the strongest when it was in what general area?

4. In which direction did the hurricane move while it was over the Atlantic Ocean?

5. What parts of the United States are most vulnerable to hurricane damage?

6. Hurricanes need energy to form and be sustained. Where does the energy come from?

7. Why do hurricanes weaken as they move over land areas? (Please state at least two reasons.)

8. How can you tell when this storm system traveled the fastest?

9. How does a hurricane differ from each of the following?

   a. Tornado _____

   b. Typhoon _____

   c. Mid-latitude cyclone _____

10. Figure 23-3 shows the path of the major Atlantic hurricanes in 2004. These paths are generally typical of hurricanes. Describe the typical path of hurricanes that come ashore in North America.

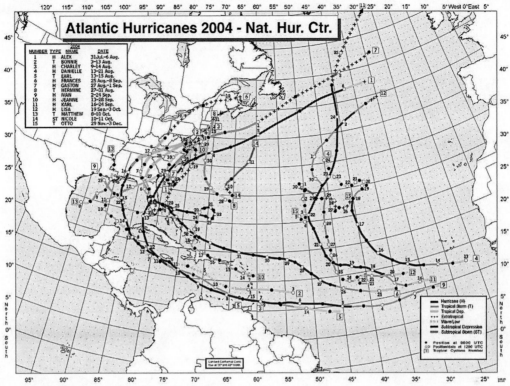

FIGURE 23-3. Hurricane tracks in 2004.

_____

_____

11. When, where, and what were the death tolls of the most deadly hurricanes (tropical storms) in the United States and worldwide? (You may need to use the library or the Internet to find the answer.)

_____

_____

12. A major hurricane can generate 1.25×10^{19} calories of energy per day. If all this energy came from condensation, how many liters of water must condense each day to form hurricane clouds and precipitation? If this were placed in a cubic container, how big would the container be?

_____

_____

_____

_____

_____

My Notes

CHAPTER 23—LAB 2: GAIA

Introduction

British scientist James Lovelock proposed the Gaia theory in the 1960s. Gaia theory proposes that Earth is a self-regulating network of interdependent physical and biological systems. The theory includes three main ideas. The first two are widely accepted.

First: Earth contains a great variety of living and nonliving things that depend upon one another. Second: Our planet exists in a delicate state of equilibrium. Changing any part of Earth can have profound effects on a variety of different Earth systems. Small stresses of limited geographic extent are likely to result in a temporary local imbalance of that equilibrium. The environment can generally recover from these small changes such as localized pollution problems. However, much larger changes applied over wide areas for long periods can result in a new equilibrium. These new conditions could be hostile to established life-forms. We can compare Earth to a rubber band. Stretch it a little, and it bounces back. However, stretch it too far, and a permanent change (in the case of a rubber band, breakage) will occur.

Lovelock's third idea is more controversial. He has suggested that our planet itself is really a gigantic living organism. This life-form is composed of organs, such as the oceans, forests, and atmosphere, with each part having a kind of biological function, such as photosynthesis, natural selection, and tectonic plate motions. However, many scientists are skeptical about taking the Gaia theory that far.

One of the most valuable aspects of the Gaia hypothesis has been a new understanding of the evolution of Earth's atmosphere. From studying other planets and from studies of very old geological formations, we know that the atmosphere of early Earth was very different from the air of today. Use the following data to show the inferred composition of the atmosphere for the 4.5 billion years of Earth's history.

Objective

In this lab, you will use the data presented to draw a graph that shows how Earth's atmosphere has changed over time.

Materials

Data, colored pencils

TABLE 23-2. Changing Percentage Composition of Earth's Atmosphere

Gas	Millions of years before the present									
	4500	**4000**	**3500**	**3000**	**2500**	**2000**	**1500**	**1000**	**500**	**Present**
Carbon Dioxide (CO_2)	80%	20%	10%	8%	5%	3%	1%	0.07%	0.04%	0.025%
Nitrogen (N_2)	10%	35%	55%	65%	72%	75%	76%	77%	78%	78%
Hydrogen (H_2)	5%	3%	1%	0.5%	0%	0%	0%	0%	0%	0%
Oxygen (O_2)	0%	0%	0%	0%	0%	1%	5%	10%	15%	21%
Others	5%	42%	34%	26%	23%	21%	18%	13%	7%	1%

Procedure

1. Figure 23-4 is a special time graph. Use the data above to construct a graph of the changing percentage composition of Earth's atmosphere since the formation of Earth. Please note that these figures are cumulative. That is, each gas must be shown on top of the previous gas, and they must add up to 100%. This is similar to the mineral composition chart of igneous rocks at the bottom of page 6 in the *Earth Science Reference Tables*. Label each gas with its chemical symbol within or next to its region on the graph. (CO_2, N_2, H_2, O_2, and other gases)

2. When you have constructed and labeled your graph, add the events listed in Table 23-3 at the proper time. Write each with an arrow showing the appropriate place below the graph. (Please note that the first arrow has been done for you on Figure 23-4.)

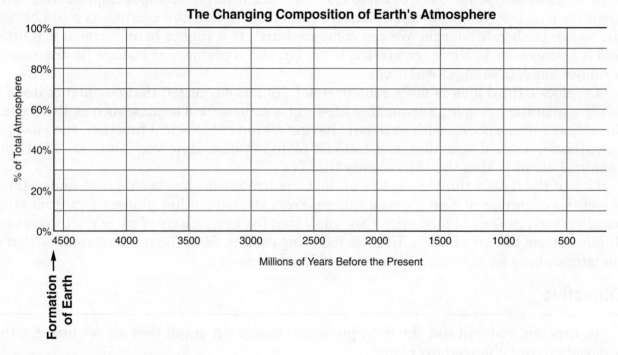

FIGURE 23-4. Blank Gaia graph.

TABLE 23-3. Selected Events in the Evolution of Planet Earth

Event	Occurrence (million years ago)
1. Formation of Earth.	4600
2. Oldest Known Bedrock	3900
3. Oldest Rocks of Organic Origin	3700
4. Precambrian Iron Deposits	3700–1800
5. Photosynthesis in Plants Begins	3000
6. Oxygen in Air Dominates Weathering	2100
7. Limestone Deposition Becomes Common	1800
8. Fossils Become Abundant	540
9. Earliest Plants and Animals on Land	420

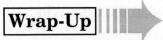

Wrap-Up

1. What is the approximate age of planet Earth?

    _____

2. How long do scientists think living things have existed on Earth?

    _____

3. How many millions are there in a billion? _____

4. From your knowledge of biological evolution, what characteristic of organisms older than 540 million years makes fossils of these life-forms relatively rare?

    _____

5. Why was oxidation (chemical weathering) not common more than 2 billion years ago? (*Hint:* See the graph.)

    _____

6. Based on your knowledge of life science, what is the primary gas is given off by green plants during photosynthesis?

    _____

7. According to your graph, what gas was depleted (used up) by the time oxygen became abundant?

    _____

8. Why was it not possible for this gas to exist with oxygen in the atmosphere?

    _____

9. What seems to have been the major cause of the dramatic change in the composition of Earth's atmosphere over the past $4\frac{1}{2}$ billion years?

    _____

10. What major atmospheric gas is increasing rapidly? What is the most likely major problem associated of this change?

    _____

    _____

    _____

11. How do these changes in the atmosphere illustrate the Gaia theory?

    _____

12. Mars and Venus have an atmosphere that is dominated by carbon dioxide. What is the major component of Earth's atmosphere?

_____

13. Aside from providing oxygen for respiration, how else did the production of elemental oxygen within the atmosphere allow the development of life-forms on land? (*Hint:* This feature is now being threatened by our technology; specifically by refrigerants like CFCs.)

_____

_____

14. When you think that Earth is billions of years old, it might be hard to understand how long that is. How tall, in miles, would a stack of 4.5 billion pennies be? Express this distance in terms that make it understandable.

_____

_____

_____

15. Why do most scientists reject Lovelock's third statement about Gaia?

_____

_____

Chapter 24
Patterns of Climate

 CHAPTER 24—SKILL SHEET 1: CLIMATE CONTROLS

Climate describes the average weather conditions over a long span of time. The difference in time distinguishes climate from weather. The climate of any region can generally be specified with three factors:

1. **Rainfall:** Rainy, humid climates are moist; arid climates are dry.
2. **Average Temperature:** Tropical climates are always warm; polar climates are usually cold; temperate climates are seasonal.
3. **Temperature Range:** Continental climates (far from oceans) have large seasonal temperature changes. Locations near large bodies of water have less significant extremes between the warmest and coolest months. Locations at high latitudes have a large range in temperatures.

Most of New York State has a mid-latitude climate, with hot summers and cold winters, and with adequate rainfall distributed throughout the year. This is a temperate, humid, continental climate.

1. What is the difference between weather and climate?

_____

2. Complete the analogy: News is to history as weather is to _____.

The following sections illustrate how geographic factors affect the climate of a region.

I. Latitude

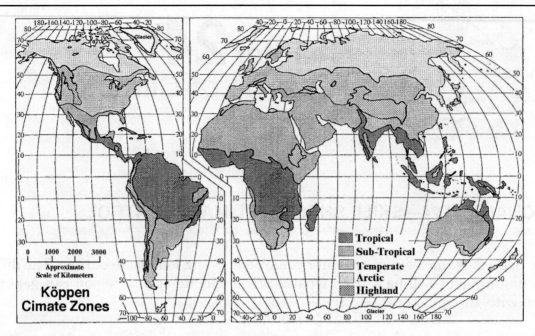

FIGURE 24-1. World climate classified by temperature zones.

3. As latitude increases, the average temperature generally _____.

4. In what part of the world does the middle of summer occur in January?

   _____

5. The temperature is always warm, so it is always like summer near the

   _____.

II. Altitude

FIGURE 24-2. Early autumn snow in the Rocky Mountains of Colorado.

6. In Figure 24-2, where is most of the snow? _____

7. Because air cools as it rises, the average temperature in high-altitude mountain regions is lower than the average temperature in nearby lowland locations. What term describes the cooling of a gas by expansion?

_____

III. Mountain Barriers

Figure 24-3 is a simplified representation of the American Pacific Coast. The coastal climate is cool and moist. Air rising over the windward side of the mountains loses its moisture as rain and snow. However, as the air descends the other side of the mountains, it is warmed by adiabatic compression. The air's temperature goes up and its relative humidity goes down, forming the deserts on the leeward (downwind) side of the mountains.

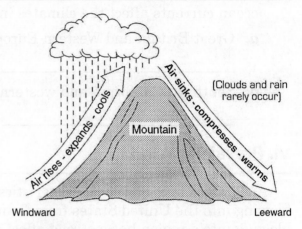

(Clouds and rain rarely occur)

Air rises - expands - cools

Air sinks - compresses - warms

Mountain

Windward   Leeward

FIGURE 24-3. Mountains affect both air temperature and precipitation.

8. On which side of a mountain range does most of the precipitation usually occur?

_____

9. Why is the climate relatively dry on the downwind (leeward) side of most mountain ranges?

_____

_____

IV. Nearness to Oceans

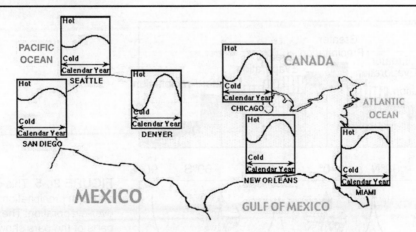

FIGURE 24-4. Coastal and inland locations have different climates.

10. According to Figure 24-4, which of these cities in the United States has the greatest temperature range (greatest change in temperature) throughout the year?

_____

11. According to Figure 24-4, how do coastal climates differ from inland climates?

_____

_____

V. Ocean Currents

12. Places near the oceans are often affected by warm and cold ocean currents. Look carefully at the ocean currents map in the *Earth Science Reference Tables.* How do ocean currents affect the climates in the following locations?

  a. Great Britain and Western Europe?

  _____

  b. Southern California and western South America?

  _____

VI. Prevailing Winds

Air masses take on the characteristics of their place of origin. For example, air masses blowing into the United States from Canada are usually relatively cold. The source of air blowing into a region has a strong affect on its climate. The Planetary Winds chart in the *Earth Science Reference Tables* shows the prevailing wind pattern around the whole Earth.

13. What is the approximate latitude of New York State? _____

14. What is the prevailing wind direction at this latitude? _____

15. In New York, do most of the winds come off the continent or off the ocean? What does this mean for the climate in New York?

_____

_____

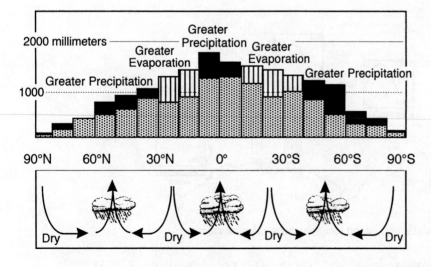

FIGURE 24-5. This graph shows where on Earth precipitation generally dominates over evaporation. The dotted and black parts of the bars show precipitation. The bottom section is a vertical profile of winds from the North Pole to the South Pole.

16. According to Figure 24-5, what part of Earth has the greatest annual precipitation?

_____

17. At what latitudes is precipitation insufficient to supply all the evaporation that takes place?

_____

18. In the latitudes where there is extra precipitation, what is the dominant vertical motion of the atmosphere? (See the bottom of the chart.)

_____

19. Why is there so little precipitation at the North and South poles?

_____

_____

20. Which dominates (precipitation or evaporation) at the latitude of New York State?

_____

21. How does each of these factors affect the local climate?

    _a._ Altitude:

    _____

    _b._ Mountain ranges:

    _____

    _c._ Nearness to water:

    _____

    _d._ Ocean currents:

    _____

    _e._ Latitude:

    _____

My Notes

CHAPTER 24—LAB 1: CLIMATES OF AN IMAGINARY CONTINENT

Introduction

When we consider the weather, we think of the conditions of the atmosphere over the past few hours. The climate of an area is the average conditions of its weather over a long (50 to 100 years) period. Climate includes the average temperature and precipitation, as well as the expected seasonal variations in these factors.

There are a variety of ways to classify climates. The system shown on the map in Figure 24-6 uses the local plant community to classify the climate. The local plant community is an especially good indicator of the climate because of the sensitivity of plants to geographic and long-range atmospheric conditions.

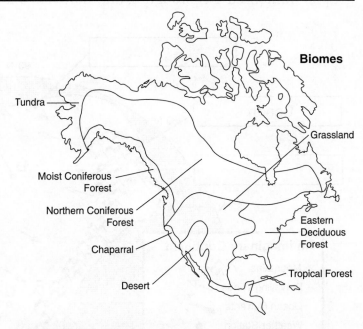

FIGURE 24-6. Climates of North America classified by plant communities.

1. What biome is used to classify the coldest climate in North America?

   _____

2. According to this map and classification system, what kind of climate does New York State have?

   _____

3. Give the common names of at least two tree species that identify the general climate of New York State.

   _____

Objective

To use geographic clues to determine the climate of areas of the Imaginary Continent

Materials

Imaginary Continent map, 8 climatographs

Procedure

Figure 24-7 is a map of an imaginary continent. Letters A through H indicate the location of eight weather stations on the continent. Each station is described by one of paragraphs numbered 1 thorough 8, and matches one of the graphs of Figure 24-8.

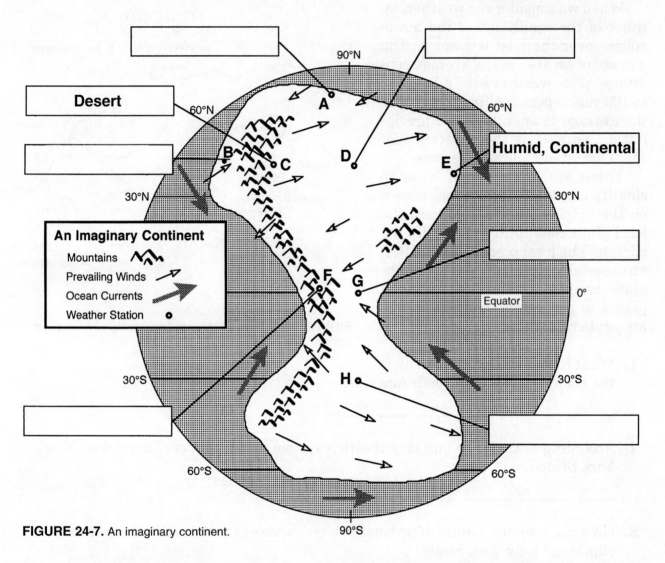

FIGURE 24-7. An imaginary continent.

A. Carefully read the numbered paragraphs called Climate Descriptions. Look at the map. You will notice weather stations on the map of the continent that are identified by letters. Each numbered paragraph describes the conditions at one of the weather stations (A–H).

B. After reading a paragraph, look at the map and choose the weather station best described by that paragraph. Write the letter of that weather station on the line provided at the end of the paragraph. (Each letter will be used only once.) If you are uncertain about which station matches, read the paragraph again. Note that the first paragraph has been labeled for you. This should help you understand how to complete the lab.

C. Using the list of eight climate types below, write the correct climate type found at each station in its box at the edge of the map.

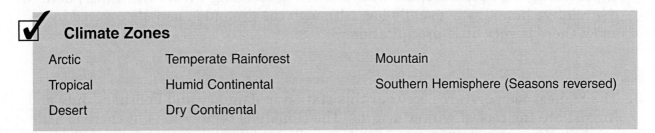

✓ **Climate Zones**

Arctic	Temperate Rainforest	Mountain
Tropical	Humid Continental	Southern Hemisphere (Seasons reversed)
Desert	Dry Continental	

D. Examine the graphs in Figure 24-8. Label each graph with the letter of the weather station that has the climate described by that graph. (Note that each graph includes both temperature and precipitation variables in a yearly cycle.)

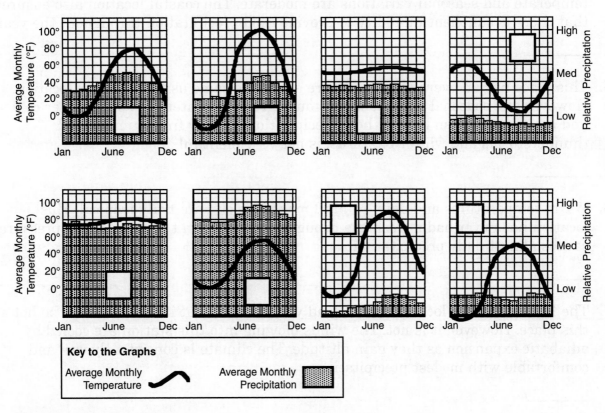

FIGURE 24-8. Eight climatographs.

Climate Descriptions

1. Although this location has a variable mid-latitude climate, the prevailing winds have crossed a major mountain range. There is little rainfall here because the air has dropped its moisture on the windward side of the mountains. Because it is far inland, there are large seasonal variations in temperature.

2. Due to its high latitude, this location receives weak insolation. The sun is never high in the sky. Winters are bitter cold, while summers are very cool. In this zone of prevailing high pressure, the wind is often descending within the atmosphere so the relative humidity is usually low. Although the ground is usually covered by snow, there is very little precipitation.

⎯⎯⎯⎯

3. The highest temperatures occur at this station in January and February. July and August are the coolest winter months. The climate is temperate, but there is little precipitation because this is a zone of prevailing high pressure and descending air.

⎯⎯⎯⎯

4. This station is influenced by a nearby warm ocean current. The climate is temperate and seasonal variations are moderate. The coastal location also ensures that there is sufficient humidity to provide good precipitation throughout the year.

⎯⎯⎯⎯

5. This place has an average temperature about the same as the previous location. However, the seasonal variations are much greater. Winters are often bitter cold and the summers can be very hot. Precipitation, largely from summer thunderstorms and winter blizzards, is not very plentiful.

⎯⎯⎯⎯

6. Warm temperatures and daily rainfall make this climate truly tropical. This location has never had a frost or a drought. Rising air in this zone of low pressure causes almost daily precipitation.

⎯⎯⎯⎯

7. The latitude of this location might lead you to think that it would always be hot at this place. However, it is not. The winds blowing into this station are cooled by adiabatic expansion as they gain altitude. The climate is consistently cool and comfortable with modest precipitation.

⎯⎯⎯⎯

8. There is lots of rainfall here because the winds blowing up the windward side of the mountains are quickly cooled below the dew point. The result is perpetual clouds and rainy weather. The ocean winds, that provide the moisture, also moderate the temperatures. The climate is temperate with warm summers and cool winters.

⎯⎯⎯⎯

Chapter 25
Earth, Sun, and Seasons

 CHAPTER 25—SKILL SHEET 1: EARTH'S SEASONS

Although the distance between Earth and the sun varies on an annual cycle, this change is too small to cause our seasons. In fact, Earth is actually closest to the sun in the month of January.

Our seasons are caused by the $23\frac{1}{2}°$ tilt of Earth's axis. This tilt causes the apparent path of the sun through the sky to change throughout the year. Summer starts when the noon sun is highest in the sky, as well as when the sun is in the sky for the greatest number of hours of the day. See Figure 25-1.

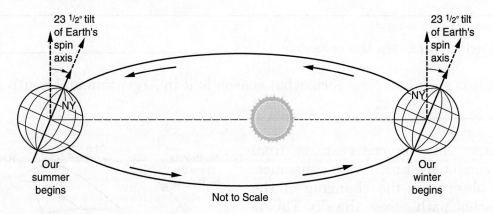

FIGURE 25-1. Earth's seasons are caused by the tilt of Earth's axis, which points to Polaris throughout the year. Note that New York is turned toward the sun in our summer.

In the Northern Hemisphere, daylight is longest about June 21. That is when our sunlight is strongest and most direct. However, south of the equator, this is when the sun is lowest in the sky and when daylight is shortest. Therefore, the seasons are six months out of phase north and south of the equator.

Figure 25-2 shows that Earth is actually about 3 percent closer to the sun in our winter. This very small change in Earth-sun distance has very little effect on our weather. If our changing distance from the sun caused the seasons, winter and summer would occur at the same time north and south of the equator. However, they do not.

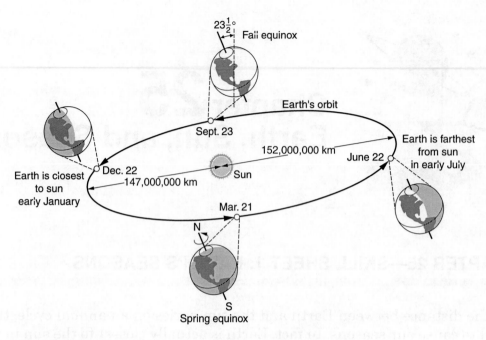

FIGURE 25-2. Earth is actually slightly closer to the sun in our winter.

1. In what month is Earth farthest from the sun? _____

2. Why does the changing distance between Earth and sun not affect the seasons?

   _____

3. What actually causes the seasons? _____

4. When it is spring in New York, what season is it in Argentina and South Africa?

   _____

Of course, in our observations from Earth we cannot see Earth from a distance. What we observe is the changing of the sun's apparent path across the sky. This is illustrated in Figure 25-3.

5. According to Figure 25-3, approximately how high in the sky is the noon sun in the New York region on June 21 _____

6. How high is the noon sun in New York on December 21? _____

7. What is the annual range (the change in altitude)? _____

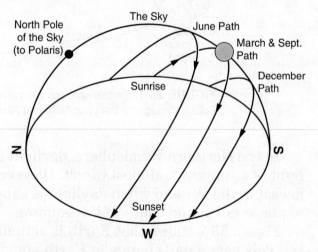

FIGURE 25-3. The apparent path of the sun through the sky changes in an annual cycle. This diagram represents observations taken in central New York State.

8. How does this range of altitude compare with Earth's tilt angle?

Each letter in Figure 25-4 represents a position of Earth in its annual revolution around the sun. Base your answers to items 9–18 on Figure 25-4.

9. What month is represented by each letter in Figure 25-4?

    A. _____

    B. _____

    C. _____

    D. _____

For items 10–18 on the blank line write the letter indicating the position described by the statement.

10. The day when the noon sun is highest in New York: _____

11. The date of the most concentrated insolation in New York: _____

12. Our first day of winter: _____

13. First day of spring in New York _____

14. Earth's approximate closest approach to the sun (perihelion): _____

15. Autumnal equinox: _____

16. The beginning of summer in the Southern Hemisphere: _____

17. Earth's greatest distance (approximate position) from the sun (aphelion): _____

18. The shortest day of the year in the United States: _____

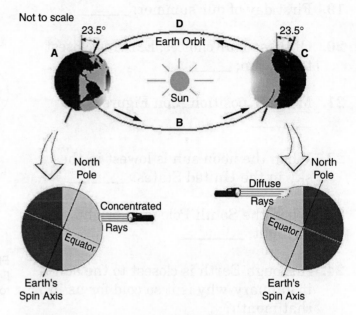

FIGURE 25-4. The two enlarged Earths show that sunlight is most concentrated where the sun is high overhead.

Base your answers to items 19–24 on Figure 25-5.

19. First day of our summer: _____

20. Of these four choices, Earth is closest to the sun: _____

21. Matches position A on Figure 25-4: _____

22. When the noon sun is lowest in the sky in the United States: _____

23. When the South Pole is in bright daylight: _____

24. Although Earth is closest to the sun in January, why is it so cold for us in that month?

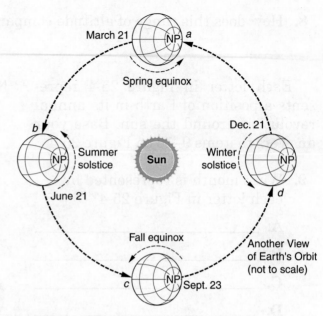

FIGURE 25-5. Earth's orbit as seen from far above. From this point of view, the orbit of Earth looks perfectly round.

🌐 CHAPTER 25—LAB 1: INQUIRY INTO SHADOW TIME

Introduction

Have you ever seen or made a sundial? How do they work? Can a sundial be used to do more than tell time? This lab should help you answer those questions?

Objective

You will be making a special graph to show changes in the length of a shadow of a vertical meterstick and a graph of the sun's changing angle over the period of one day. Your graphs should extend from sunrise to sunset.

Materials

You will need a large and level outside surface on which to measure the shadow. A level driveway or large concrete surface is best. (You can combine measurements taken at different places, such as home, and school.) You also need one or more long narrow objects such as a meterstick (39.4 inches long). (Any straight stick of about this length will work.) You will need something to measure metric length such as a second meterstick or ruler as well as a protractor, the larger the better. A scientific calculator is also useful.

Procedure

Design the procedure you will use to measure throughout the day the length of the shadow cast by a 1-meter-long object. Have your procedure approved by your teacher before you begin.

Please hand in the following:

1. A data table of shadow length and sun angle measured four to eight times from sunrise to sunset. Please construct and submit your data table on a separate sheet of paper. Data from different days can be combined as long as they were not made more than a week apart.

2. Two graphs (sunrise to sunset) with the axes properly labeled. One will show the length of the shadow and the second will show the angle of the sun. On each graph add two vertical lines, top to bottom, to show the times of sunrise and sunset.

3. The answers to questions 1–12 below.

You will be graded on the basis of accuracy, readability, neatness, and application of the skills that you learn. (*Hint:* Think carefully about how you would expect the shadow length to change during the day.)

Wrap-Up

1. Briefly describe how you obtained the shadow length data in your data table

   _____

   _____

2. When could a shadow first appear? _____

3. According to your graph, when was the shadow longest? (two answers)

   _____ and _____

4. Briefly describe the changing length of the shadow from sunrise to sunset, as shown on your graph.

   _____

   _____

5. Why is the path of the sun through the sky considered an apparent motion, and not a real motion?

   _____

6. Where must the sun be if the length of the shadow were to shrink to 0 meters?

   _____

7. Can a vertical object have shadow length of zero in the continental United States? Explain.

   _____

8. How does the length of the noon shadow change throughout the year at your latitude? (Please describe the change.)

   _____

9. For an observer in New York State, in what general compass direction is the sun at sunrise? (North, South, East, or West)

   _____

10. What is the general direction of the sun at sunset? _____

11. In what part of the sky is the noon sun in New York? (North, South, East, or West)

    _____

12. In what compass direction will the noon sun at any location in New York State cast a shadow?

    _____

CHAPTER 25—LAB 2: PATHS OF THE SUN

Introduction

Perhaps the most dependable cycles in our lives are the daily motion of the sun across the sky and the changing of the seasons. These changes are a result of the tilt of Earth's axis along with the rotation and revolution of Earth. These motions are remarkably constant. Even the great Sumatra earthquake of 2004 that generated the disastrous tsunamis in the Indian Ocean changed Earth's rotation by only about one part in a billion.

Your teacher will show you a plastic hemisphere that represents the sky. On the outside surface of the hemisphere you can show the positions of the sun, the stars, and any other celestial objects. However, this model can also be used to illustrate motion. You can draw lines to represent the apparent motions of these objects, as they seem to move through the sky.

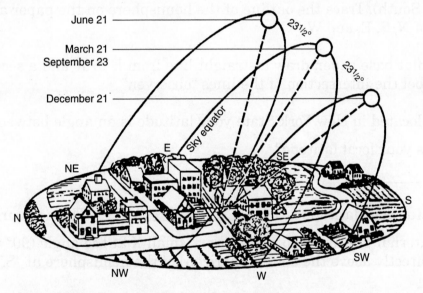

FIGURE 25-6. This model of the sky is drawn from the west.

Objective

In Chapter 1, you probably constructed a model to visualize the size and shape of a part of Earth's surface. You tried to make this model as accurate as you reasonably could to represent the true landscape. In this chapter you will be constructing physical or imaginary models to help you understand the scale, position, and motions of Earth.

Materials

Clear plastic hemisphere, ruler, external protractor, cardboard

Procedure

These directions may be difficult to follow. It may help to have one lab group member read the directions aloud step by step. Another person can help interpret the directions. The third can actually perform these steps. Be sure you all agree on what to do at each step.

1. Obtain one clear plastic hemisphere, a ruler, and an external protractor for your lab group. (Be sure to conform to the maximum group size specified by your teacher.)

2. Place the hemisphere on a table or desk with the convex side up. Place paper under the hemisphere to avoid marking on the desk. Use a water-soluble thin-line marker to draw a thin, dark vertical line about 1 cm long from the table upward on the hemisphere. Label that line N (for North). Then, 90° clockwise, make a similar vertical line up from the edge and label it E (East), 90° counterclockwise from N make a third line and label it W (West). The final vertical line at 90° should be labeled S (South). Trace the outline of the hemisphere on the paper and label the positions of N, S, E, and W.

3. Along the flat base paper, draw a straight line from E to W and a second line from N to S. Label the intersection of the lines "observer."

4. If you are located in New York State your latitude is an angle between 41° and 45°

   a. What is your local latitude?

   _____

   b. 90° minus your latitude is _____. This is your sun angle.

   Use the external protractor to place a very small X at the angle (90° minus your latitude) directly above the bottom (horizon) of the hemisphere at "S."

5. Lay your external protractor along the hemisphere so the 45° mark is at the X you made in the last step. Then align the protractor so the two ends point to the E and W marks. The protractor should be in a plane that intersects the paper at the observer position and at E and W. (So the plane of the protractor will slope onto the desk.) Trace a line through the X and along the edge of the protractor.

6. Now connect both ends of the line to the E and W positions at the horizon level. This should make a straight line with no angles. Label this line "Sun's path at the equinoxes."

7. Make several marks $23\frac{1}{2}°$ higher than the first sun path. You must measure perpendicular to the first line you drew so the sun paths are parallel. Connect those points to make a second and higher and longer sun path. Label it "June solstice." Make a third and shorter sun path $23\frac{1}{2}°$ lower than the first line and label it "December solstice."

           LABORATORY MANUAL

8. Where on this hemisphere do you think the sun path will be today?

_____

_____

9. Now you will use the method shown in Figure 25-7 to show the path of the sun across the sky today. Place the same hemisphere on a flat outdoor surface with the compass directions (N, S, E, and W) in their true directions. (If possible, determine North from previous observations of Polaris.) Use a water-soluble marker to make a dot on the plastic so that the dot casts a shadow exactly at the position of the observer as in Figure 25-7. Label this dot with the clock time to the nearest minute.

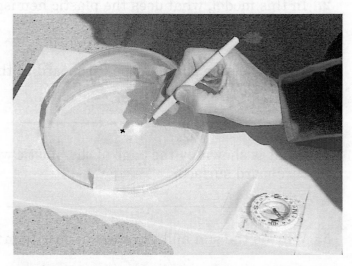

FIGURE 25-7. Recording the path of the sun through the sky.

10. Repeat this procedure at various times of the day separated by at least an hour or so. (You may be able to do this with students in earlier and later Earth science classes.) The dots should make a path parallel to the other paths you have drawn. Connect the dots to make the full path of the sun today.

11. Use the external protractor to measure the length of each sun path in degrees of angle. Enter all data in Table 25-1. Some boxes have been filled in to show you what is expected. The fifth column should be the angle of the noon sun above the southern horizon. The hours of daylight will be the total sun path angle divided by 15°/hour.

TABLE 25-1. Paths of the Sun at Your Latitude

Celestial Event	Date/Month	Sunrise Direction	Sunset Direction	Noon Sun Altitude	Total Path (degrees of angle)	Hours of Daylight
Spring equinox		Due East			180°	12 hours
		Northeast	Northwest			
			Southwest			

Questions About the Model

1. What is a model? _____

2. In this model, what does the plastic hemisphere represent?

   _____

3. What does the point on the cardboard at the center of the plastic hemisphere in Figure 25-7 represent?

   _____

4. What is shown by the edge of the plastic where the hemisphere meets the cardboard surface?

   _____

5. Why do the sun and the stars appear to move through the sky?

   _____

6. Is this apparent motion a cyclic or a noncyclic change? _____
   Why? _____
   _____

 Wrap-Up

1. What is the tilt of Earth's axis with respect to a line perpendicular to Earth's orbit?

   _____

2. In the United States, during what month is the noon sun highest in the sky?

   _____

3. At what time of year is the noon sun directly overhead (at the zenith) in New York State? (Hint: answer carefully.)

   _____

4. What is the altitude angle of the zenith for any location on Earth? _____

5. Why is the path of the sun an apparent motion and not a real motion?

   _____

6. In what two ways does the path of the sun through the sky change from late December to late June?

   _____

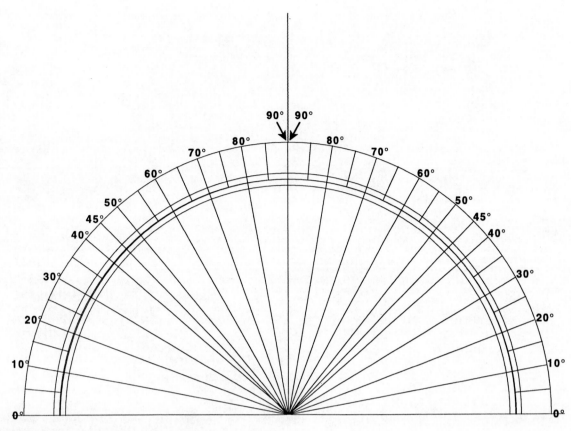

FIGURE 25-7. External protractor template

7. In what two ways does the path of the sun through the sky change from late June to late December?

_____

8. In the Northern Hemisphere, why is the length of daylight longer in the summer than in winter?

_____

9. In the Northern Hemisphere, during which two seasons does the point of sunrise move to the south?

_____ and _____

10. When does the sunrise point move northward in the Northern Hemisphere?

_____

11. From our observations of the sky, why is sunlight stronger and last longer during the summer than during the winter?

_____

12. What feature of Earth's orientation causes our seasons?

_____

13. What is the yearly pattern of daylight and darkness from January through December at the North Pole?

   _____

   At the South Pole?

   _____

14. When is Earth actually closest to the sun?

   _____

15. Although the sun is never at the zenith in New York State, in what part of Earth's surface is the noon sun located at the zenith at least one day of the year?

   _____

16. You have learned that day and night are caused by the rotation of Earth. Suppose someone asked you to prove by an experiment that Earth is spinning, that is, to prove that we are not observing the sun orbiting Earth, How could you prove it? (Citing books or anyone's word is not a scientific experiment.)

   _____

   _____

CHAPTER 25 SKILL SHEET 2: THE VERTICAL RAY

The sun is always at the zenith at some place on Earth. The vertical ray is the ray of sunlight that comes from the zenith, striking Earth's surface at an angle of 90°. The vertical ray appears where the sun is directly overhead.

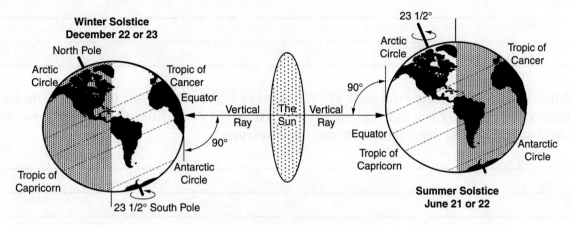

FIGURE 25-9. Earth's orientation at the solstices. Like most other diagrams in this chapter, this is not to scale.

1. The vertical ray strikes Earth's surface at an angle of _____

2. About June 21, the vertical ray strikes the _____

3. The vertical ray hits the Tropic of Capricorn in _____

4. On June 21, the sun is visible at midnight north of the _____

5. When does the vertical ray strike the Tropic of Cancer? _____

6. The South Pole is in total darkness all day on what date? _____

7. The word *equinox* is made up of two parts. Can you guess what each one means?

   equi- means _____ and

   -nox means _____.

8. What is the latitude of the Tropic of Cancer?

   _____

   The Tropic of Capricorn? _____

   The North Pole? _____

   The Arctic Circle? _____

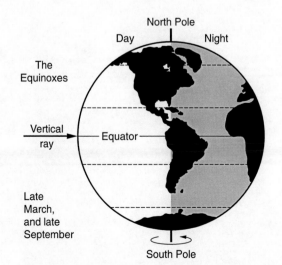

FIGURE 25-10. Earth at the equinoxes.

9. How could you determine the exact position of the Tropic of Cancer without using a map or any previous knowledge of its position?

   _____

10. How would the seasons be affected if Earth's tilt were to increase to an angle significantly greater than $23\frac{1}{2}°$?

   _____

   _____

   _____

11. Albert Einstein postulated that gravity makes light rays seem to bend. This has been confirmed by scientific observations. Einstein explained that gravity causes the "warping of space." How can this be observed?

   _____

   _____

   _____

   _____

CHAPTER 25 SKILL SHEET 3: THE CHARIOT OF HELIOS

According to Greek mythology, Helios (the god of the sun) was responsible for driving the horses that pulled his fiery chariot (the sun) across the sky. In that job, not showing up at sunrise just was not an option.

Use the clues below to complete the crossword puzzle in Figure 25-11.

FIGURE 25-11.

ACROSS

1. Appears to be true
5. Before noon
7. Month of spring equinox
9. Compass direction of noon sun
10. "Life _____ the Mississippi"
12. The most powerful playing card
14. How smart you are
16. Direction of the zenith
18. Not difficult (shortened phonetic form)
20. In the summer, the noon sun is _____ in the sky
21. Direction of Polaris
23. Straight overhead
25. The time each day when the sun is highest in the sky
27. Our country (abbr.)
30. A strong barrier or fence
32. English beverage consumed at 4 P.M.
33. Longest and shortest days of the year.
35. Azimuth or altitude
36. Homonym of 2
37. Nickname for a man from Texas
38. _____ "the twilight's last gleaming"
39. Rooster's way to announce sunrise
40. In the summer, the noon sun is _____ in the sky
41. The pole that has continuous night from March to September
42. The sun is the nearest _____

DOWN

1. The angle of direction
2. Ma's husband
3. Dates of equal day and night
4. Homonym of "too"
6. Graduate degree after Bachelor of Arts
7. He/him : I/_____
8. Liquid precipitation
9. Earth's primary energy source
11. Sunrise direction in the summer (abbr.)
13. On June 21, the vertical ray strikes the Tropic of _____
15. Month of the winter solstice (abbr.)
17. A small boat used during World War II
19. The highest point in the sky
20. The sun and stars appear to move 15°/_____
22. $23\frac{1}{2}°$ is the _____ of Earth's axis
24. She/he : hers/_____
26. Container to carry groceries home
28. Sunrise direction on winter solstice (abbr.)
29. Direction in which stars seem to rise
30. On September 21, the sun sets due _____
31. In winter, the noon sun is _____ in the sky
33. Repair a seam
34. Children's game
37. Antonym of "from"

Challenges

1. In Shakespeare's play *Hamlet,* young Hamlet makes a pun about the sun. What is his statement? Why is it ironic?

2. One way to determine an upper limit for the age of the sun is to discover how it generates heat and light (energy). About 150 years ago Lord Kelvin calculated that if the sun were made of coal, it would burn out in just 800 years. (In his day, nuclear energy was not known.) However, like Earth, the sun is about 5 billion years old. So how many times as much energy per ton comes from nuclear fusion than from burning coal?

CHAPTER 25—LAB 3: INQUIRY INTO THE APPARENT MOTION OF THE SUN

Introduction

The sun moving through the sky is one of humankind's most dependable events. Ancient cultures proposed many explanations of how and why the sun moves through the sky. However, most of these explanations involved the sun actually moving. Scientists understand that the sun does not move around Earth. It is the revolution of Earth on its axis causes the apparent motion of the sun.

Materials

If your classroom faces the sun, you can pass sunlight through a pinhole. (The room must be very dark.) If the classroom has windows that face away from the sun, you can use a tiny piece of a front-sided mirror. This kind of mirror has the reflective material on the front surface. Figure 25-12 shows how the front-sided mirror must be positioned.

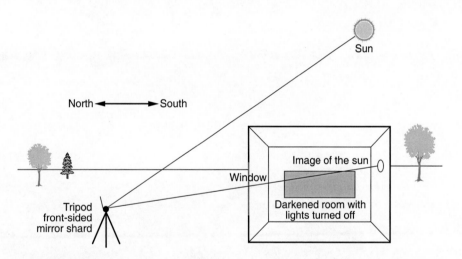

FIGURE 25-12. Projecting an image of the sun using a piece of a front-sided mirror.

Objective

To measure the apparent motion of the sun using a pinhole or a front-sided mirror.

Procedure

Devise a method to measure the sun's apparent motion. Before you begin, have your procedure approved by your teacher.

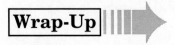

Your lab report should include your objective, materials, procedure, and conclusion(s).

My Notes

Chapter 26
Earth and Its Moon

CHAPTER 26—SKILL SHEET 1: LOADS OF LUNACY

The moon has long been associated with strange powers. Writers, painters, and composers have noted its beauty and speculated about its effects on people. In fact, the words *lunacy* and *lunatic* were used to describe people who became mentally unbalanced because they spent too much time in moonlight. The connection is no longer accepted. However, the word persists. Today, we know that the moon is just our nearest neighbor in space and the only celestial object actually visited by humans

Figure 26-1 shows eight positions of the moon as it revolves around Earth. With respect to Earth, from where are we observing this? _____

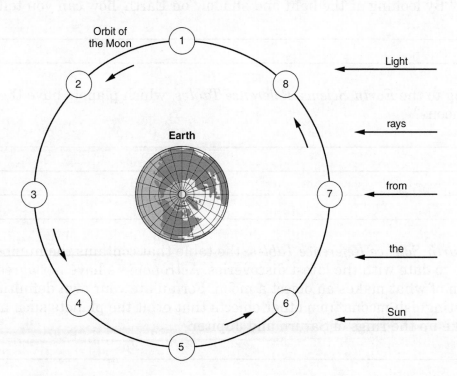

FIGURE 26-1. Moon's orbit.

Use a pencil to shade in the dark side of the moon in each of these eight moon positions in Figure 26-1. Then shade the circles in Figure 26-2 to show how the moon would appear *from Earth* at each position in its orbit.

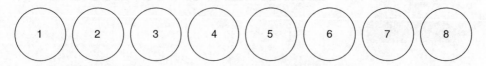

FIGURE 26-2. What do the lunar phases look like from Earth?

1. What motion of the moon causes the phases we observe from Earth?

    _____

2. Figure 26-1 is partly to scale. What single change would make the whole diagram drawn to a single scale?

    _____

    _____

3. Figure 26-1 represents a specific time of year. What month(s) is (are) shown in the diagram? By looking at the light and shadow on Earth, how can you tell?

    _____

    _____

4. According to the *Earth Science Reference Tables,* which planets have the most and fewest moons?

    _____

    _____

    _____

    _____

5. In the *Earth Science Reference Tables,* the table that contains the number of moons is not up to date with the latest discoveries. Astronomers have not agreed on a definition of what makes an object a moon. Formulate your own definition to help them distinguish moons from other objects that orbit the planets such as the rocks that make up the rings of Saturn and Jupiter.

    _____

    _____

    _____

6. Figure 26-3 shows that the sun and moon seem to change in angular diameter as we observe them in the sky. The sun and moon appear about the same size in the sky, roughly 0.5° (30 minutes) of angle. Yet the actual diameter of the sun is 400 times larger than the diameter of the moon. How can the moon look about the same size as an object 400 times its diameter?

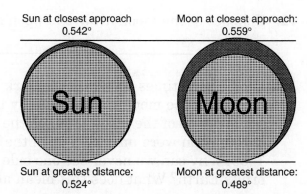

Sun at closest approach
0.542°

Moon at closest approach:
0.559°

Sun at greatest distance:
0.524°

Moon at greatest distance:
0.489°

FIGURE 26-3. The sun and moon appear about the same size in the sky. Furthermore, both of them change in apparent angular diameter, although the moon changes more.

_____

_____

7. Which appears to change more in angular size, the sun or the moon?

_____

8. What causes these objects to change in apparent angular diameter?

_____

9. Why does the moon seem to change more?

_____

_____

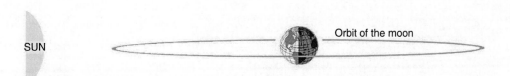

SUN   Orbit of the moon

FIGURE 26-4. The moon's orbit is inclined about 6° with respect to Earth's orbit.

10. Eclipses are among the most dramatic celestial events. Figure 26-4 shows Earth and the sun. On Figure 26-4, draw two circles to indicate the position of the moon during an eclipse of the sun and an eclipse of the moon. Draw the circles to the correct scale with respect to the drawing of Earth. Label the moon where it creates a solar eclipse "Solar Eclipse." Label the other moon position "Lunar Eclipse." Then, draw in Earth's shadow extending into space.

11. Eclipses of the sun and moon occur only at certain phases of the moon. In what phase must the moon be when an eclipse of the sun occurs.

_____

12. Eclipses of the moon always happen at what moon phase?

_____

13. People sometimes speak of "the dark side of the moon" when they really mean the far side of the moon that is actually in sunlight 50% of the time. Still, you never see one side of the moon, even as the moon orbits Earth. The features of that part of the moon were unknown until the Russians sent a camera to orbit the moon in 1959. Why can we never see one side of the moon from Earth, even as the moon orbits Earth? What does this mean about the rotation and revolution of the moon?

_____

_____

_____

14. Eclipses of the sun and moon are relatively rare. Why do these eclipses not occur each month?

_____

_____

_____

☀ CHAPTER 26—SKILL SHEET 2: WEIGHTY PROBLEMS

1. Define gravity. _____

_____

2. Figure 26-5 shows Selene standing on Earth. According to the diagram, what is Selene's normal weight?

_____

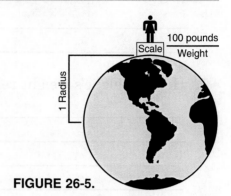

FIGURE 26-5.

Let us investigate what causes our weight.

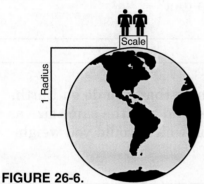

FIGURE 26-6.

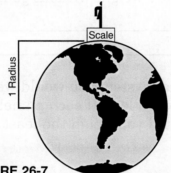

FIGURE 26-7.

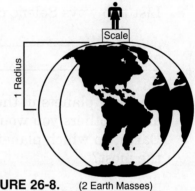

FIGURE 26-8. (2 Earth Masses)

3. Write the correct weight under the conditions shown in Figures 26-6–26-8.

26-6 _____, 26-7 _____, and 26-8 _____

4. Figures 26-5 through 26-8 show that weight is directly related to what?

_____

5. Now report Selene's weight as illustrated in Figures 26-9 and 26-10.

Figure 26-9 _____, Figure 26-10 _____

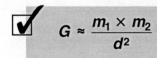

$$G \approx \frac{m_1 \times m_2}{d^2}$$

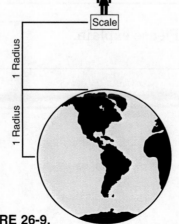

FIGURE 26-9.

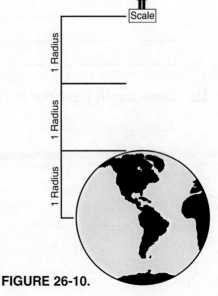

FIGURE 26-10.

6. State in words the full meaning of the mathematical formula printed in the box on page 303.

_____

_____

7. How is Selene's weight related to her distance from the center of Earth?

_____

_____

_____

_____

8. List two ways Selene can lose weight without going on a diet?

_____

_____

9. On some planets in the solar system you would weigh more than you do on Earth, and on others you would weigh less. If each planet were solid and the same size as Earth, on which planet would you weigh the least and on which would you weigh the most?

_____

10. Question 9 is based on two incorrect assumptions about the planets other than Earth. One assumption is stated in the question. What feature of Jupiter would make this comparison unrealistic or difficult to conduct as an experiment?

_____

_____

11. How are mass and weight different?

_____

_____

_____

12. Does Earth have weight? Please explain.

_____

CHAPTER 26—LAB 1: GRAVITY AND THE TIDES

Introduction

The cyclic changes in water level we know as the tides can cause river currents to change direction. The Native Americans called the Hudson River "the river that flows in two directions." In some rivers the reversal is sudden enough to make wave called a tidal bore, which sweeps upstream. The Amazon River in South America and the port of Hangzhou in China have major tidal bores. Several rivers in Atlantic Canada also have dramatic tidal bores.

FIGURE 26-11. The extreme tides in Canada's Bay of Fundy cause tidal bores in many rivers that flow into the bay. This is the Salmon River at Truro. Normal river flow is away from the observers.

Although the sun is much larger than the moon, the moon is closer to Earth, so it exerts more gravitational force. When the sun and moon are in line they pull together, the highest (spring) tides occur. This also occurs when the sun and moon are on opposite sides of Earth. This is also when the low tides are lowest. Smaller ranges of tides (neap tides) occur when the sun and moon are at right angles.

1. Which diagram on page 306 shows the positions of Earth, sun, and moon when the tidal range is the greatest?

   _____

   Which diagram shows the positions of these celestial objects when the smallest (neap) tides occur?

   _____

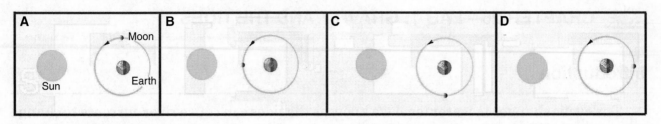

FIGURE 26-12.

2. What is the phase of the moon when the sun, moon, and Earth are positioned as shown in position A?

_____

In position B? _____

In position C? _____

In position D? _____

3. Why can a solar eclipse occur only at the new moon phase?

_____

_____

4. During what two phases of the moon do spring tides occur?

_____ and _____

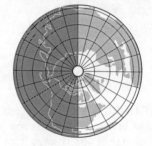

FIGURE 26-13.

5. Figure 26-13 shows the moon and Earth. On Figure 26-13, show where the tides are high by drawing a gray or blue border around Earth. The border should be thicker where the tidal rage is highest.

6. The moon is Earth's only natural satellite. What is a satellite?

_____

7. In the 1970s, American astronauts brought nearly 400 kg of rocks back from our moon during the Apollo space missions. If you had the opportunity to examine these rocks, name three minerals you would expect to find.

_____

_____

_____

Copyright © 2007 AMSCO School Publications, Inc.

Chapter 27
The Solar System

 CHAPTER 27—LAB 1: ORBITS, ELLIPSES, AND ECCENTRICITY

Introduction

Ancient observers of the night sky noticed that some of the stars did not move through the sky with the other stars. They called these objects the wanderers. Our modern word *planet* comes to us from the Greek word for a person who wanders from place to place.

In 1609 Johannes Kepler, based on years of careful observations and measurements, plotted the orbits of the five planets known in his time. Kepler showed that the orbits of the planets are not perfect circles; in fact, they are ellipses. Kepler found a way to characterize elliptical orbits with mathematical precision.

Objective

To construct ellipses and calculate eccentricity.

FIGURE 27-1. Johannes Kepler.

Materials

Approximately 30–40 cm of string, soft board about $8\frac{1}{2} \times 11''$, paper, 2 straight pins, metric ruler, pencil, safety compass

Procedure

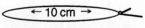

FIGURE 27-2. Ten-centimeter loop of string.

1. Tie the string into a loop 10 cm long when stretched out as shown in Figure 27-2 (The loop must be ± 0.5 cm, that is, between 9.5 and 10.5 cm.)

2. Locate the center of the sheet of paper by drawing two very light diagonal lines (in pencil) from corner to corner as in Figure 27-3. (You will need one sheet of paper per person.)

FIGURE 27-3. Finding the center of the paper.

3. Push the two pins through the paper and into the board as shown in Figure 27-4. (Do not push them all the way in.) Each pin should be located 4.5 cm from the center of the paper on a line passing through the center. (The two pins will be 9 cm apart.)

FIGURE 27-4. How to draw an ellipse.

Each pin will act as a focus (FO-cus) of the ellipse. The two pins are located at the foci (FO-sigh). Every ellipse has two foci.

4. Stretch the string around the bottom of the pins and use your pencil to draw the ellipse as shown in Figure 27-4.

In the next step, you will calculate the eccentricity of this ellipse using the equation given in the *Earth Science Reference Tables*. Eccentricity is a measure of the flatness of an ellipse. A circle is completely round, so a circle has an eccentricity of 0. A line, which is totally flat, has an eccentricity of 1. Therefore, eccentricity varies between 0 and 1, depending on the shape of the ellipse. Eccentricity is a ratio; therefore, it has no units.

1. Kepler found that the shape of all orbits are not perfect circles, but are

_____.

Each pin is located at one _____ of the ellipse. The _____ is a measure of the flatness of an ellipse. Eccentricity is always a value between _____ and _____.

2. Copy the equation for eccentricity from the *Earth Science Reference Tables* into the box below.

3. Calculate eccentricity to two significant figures. Show your calculation for the first ellipse you drew in the box below. Remember to start each calculation with the algebraic equation.

4. The ellipse you drew represents the eccentric shape of the orbits of some comets. Most planets have more circular orbits. Label this ellipse with its eccentricity. Along the ellipse line write e =

5. On the same page as your first ellipse, draw a second ellipse. This time make the distance between the two pins (the two foci) 6 cm. Each pin should be 3 cm from the center of the paper. Use the same loop of string for all your ellipses.

6. Calculate the eccentricity of your second ellipse. Show your work here as you did in step 3 above. Then label the second ellipse with its eccentricity ratio.

7. The third ellipse will show the true shape of Earth's orbit around the sun. Use the same side of the paper. This time put the pins 0.4 cm (4 mm) apart, 2 mm on either side of the center mark.

8. Label the third ellipse "Orbit of Earth." Then label either one of the two pin positions "Sun." Please note that the sun is not at the exact center of Earth's orbit. It is at one focus, which is why the sun's apparent diameter changes during the year. The other focus is just a point in space near the sun. Also note that the orbit of Earth is nearly a perfect circle.

9. Use a drawing compass to draw a circle over your ellipse that represents Earth's orbit. Does Earth's orbit come close to being a perfect circle, or is it very eccentric?

_____

10. Calculate the eccentricity of Earth's orbit around the sun in box below. Then label the drawing with its eccentricity ratio.

You have probably noticed that objects that are closer to you look larger than the same size object that is farther away. See Figure 27-5.

Throughout the year, the distance between Earth and the sun changes by about 2 percent. As seen from Earth, this causes the apparent size (angular diameter) of the sun to change (see Figure 27-6). The sun actually looks the largest in early January, when we

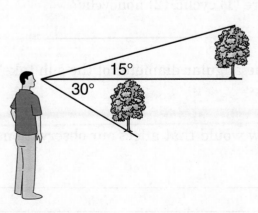

FIGURE 27-5. Angular diameter depends on size and distance.

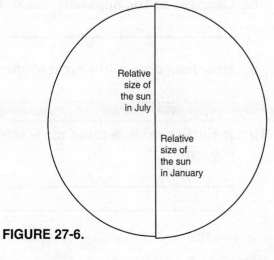

FIGURE 27-6.

LAB 1: ORBITS, ELLIPSES, AND ECCENTRICITY   **309**

are the closest to it. The sun appears the smallest in July, when the distance between Earth and the sun is the greatest. However, this change in distance is too small to affect our seasons, which are caused by the tilt of Earth's axis.

As seen from Earth, the sun and the moon have an angular diameter of about 0.5° of angle. Figure 27-7 shows an angle of about 0.5°, or 30 minutes.

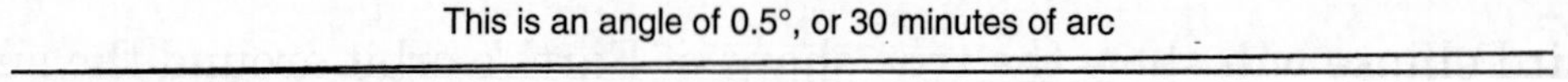

This is an angle of 0.5°, or 30 minutes of arc

FIGURE 27-7.

Wrap-Up ▌▌▌➡

1. Kepler discovered that orbits are not circles, they are _____.

2. Is the sun at the exact center of Earth's orbit? _____

3. The sun is located at one _____ of the ellipse. (The ellipse you drew should show this.)

4. Which of the following shapes best describes Earth's orbit to scale?

   football, egg, or bowling ball _____

5. What is the approximate angular diameter of the sun and the moon?

   _____

6. If Earth's orbit around the sun is an ellipse, not a perfect circle, how does this affect our observations of the sun from planet Earth?

   _____

7. In what month does the sun appear the largest to us?

   _____

8. Changes in the apparent size of the sun are (1) cyclic, (2) noncyclic?

   _____

9. How long does a full cycle of changes in the angular diameter of the sun take?

   _____

10. If Earth's orbit became more eccentric, how would that affect our observations of the sun?

    _____

    _____

11. Accurately describe the shape of Earth's orbit around the sun.

12. As Earth orbits the sun our position in space changes. During one night of observations, why do we not see the night stars change their relative positions?

13. On a separate piece of paper, calculate the eccentricities of the ellipses shown in Figure 27-8. Make your measurements to the nearest whole millimeter and carry out your calculations to two significant figures. Show your work, including the algebraic formula, for each.

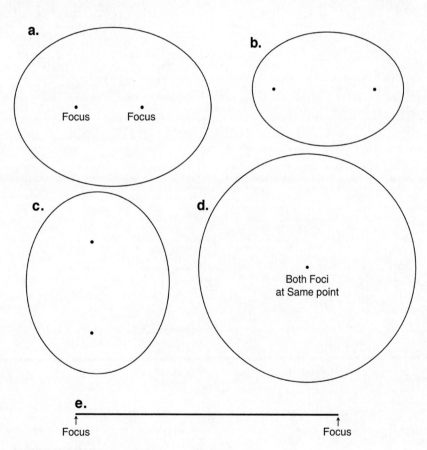

FIGURE 27-8. Five ellipses.

14. On another sheet of paper, draw ellipses with the following eccentricities

    a.   e = 0.25

    b.   e = 0.5

    c.   e = 0.95

    d.   e = 0.99

15. Why is it impossible to use your string to draw an ellipse with an eccentricity greater than 1?

CHAPTER 27—LAB 2: KEPLER'S LAWS

Introduction

Johannes Kepler was a brilliant astronomer and mathematician. Actually, Kepler was more interested in predicting the future based upon the positions of the stars (astrology) than he was in astronomy (the study of the motions of the these objects). Nevertheless, he needed to know and be able to predict the precise position of celestial objects to do his astrology. In spite of his objectives in celestial fantasy (for which he was actually employed), Kepler is best known today for his three precise and mathematical laws of planetary motion.

Objective

To discover the properties of ellipses.

Materials

None

Procedure

✓ Kepler's First Law

The orbit of a planet about a star is an ellipse with the star at one focus.

The planets are satellites of the sun, which is their primary. Earth is the primary of the moon, but Earth is also a satellite of the sun. Figure 27-9 represents a satellite orbiting

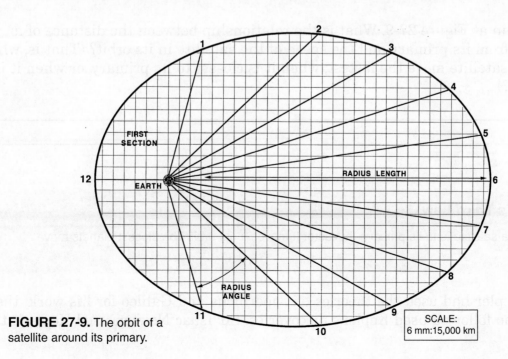

FIGURE 27-9. The orbit of a satellite around its primary.

SCALE:
6 mm:15,000 km

Earth. Each number shows the position of the satellite on successive days. How long does it take this satellite to orbit Earth? _____

The next section will help you discover how the areas of the 12 sectors compare. Count the number of squares in each section of the orbit. Count each square for which half the square is inside the section you are counting. Record your data in Table 27-1.

TABLE 27-1. Relative Area Swept Out by the Satellite Each Day.

Section	12–1	1–2	2–3	3–4	4–5	5–6	6–7	7–8	8–9	9–10	10–11	11–12
Number of squares												

1. Are the number of squares (the area) approximately the same in each section?

   _____

 Kepler's Second Law

The radius between a satellite (such as a planet) and its primary (such as the sun) sweeps out equal areas in equal times.

That is why each section of the orbit in Figure 27-9 had about the same number of little squares (area).

2. The Solar System Data Table in the *Earth Science Reference Tables* lists the mean distance of each planet from the sun. The same table gives the planets' period of rotation. Construct a graph to show how the time needed for one orbital circuit relates to the average distance of each planet from the sun.

3. Look again at Figure 27-9. What is the relationship between the distance of a satellite from its primary and the speed of the satellite in its orbit? (That is, when does the satellite move the fastest, when it is closest to its primary, or when it is far away?)

   _____

   _____

 Kepler's Third Law

The closer a satellite is to its primary, the faster it moves in its orbit. The closest planets move the fastest.

Just as Kepler had used the theories of Copernicus and Galileo for his work, the astronomers who followed used Kepler's discoveries. Sir Isaac Newton used Kepler's three laws in developing his theory of gravitation, and later Albert Einstein used Kepler's laws in his theories of relativity. Sir Isaac Newton wrote to his rival, Robert Hooke, "If I have seen farther than others, it is because I was standing on the shoulders of giants."

4. The shape of Earth's orbit is a(an) _____ with the sun located at one _____.

5. Earth is a _____ of the sun. What is Earth's major satellite? _____.

6. Earth is the closest to the sun in the month of _____.

7. Earth travels in its orbit most slowly in the month of _____.

8. In equal times, the radius to a planet sweeps out _____.

9. Which planets travel faster in their orbits, the inner, or outer planets?

_____

10. Label each of the following as a cyclical or a noncyclical change:

    a. the changing distance between Earth and the sun _____

    b. the changing speed of Earth in its orbit _____

    c. the changing angular diameter of the sun _____

11. What length of time is required for one complete cycle of the changes listed above?

_____

12. The planets can be separated into two groups based on their sizes and densities: the terrestrial (rocky) planets and the gas giants.

    a. Which planets are "terrestrial"? _____

_____

_____

    b. Which are gas giants? _____

13. *a.* Which of the gas giants has the greatest density? _____

    b. What is its density? _____

    c. Which of the terrestrial planets has the lowest density?

_____

    d. What is its density? _____

14. According to Newton, what force holds all satellites in their orbit around their primary?

_____

15. Is the orbital period of a planet affected by its mass? How do you know?

_____

_____

CHAPTER 27—LAB 3: INQUIRY INTO THE SOLAR SYSTEM TO SCALE

Introduction

If a model of the solar system were drawn to a single scale and the diameter of Pluto were 1 mm, Pluto would be 3 km from the model sun. At that same scale the model sun would be nearly 1 meter across. Consider these dimensions when you make your model. You may wish to have one scale for the diameter of the celestial objects and another for the distances between them.

Objective

Use data from the *Earth Science Reference Tables* to construct a scale model of the solar system.

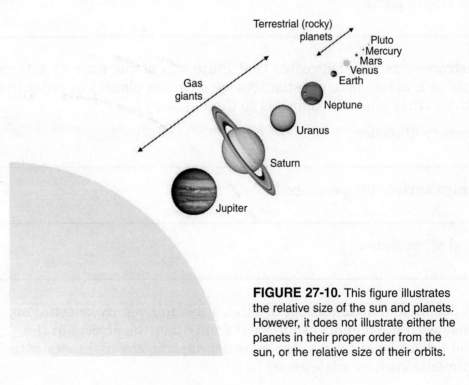

FIGURE 27-10. This figure illustrates the relative size of the sun and planets. However, it does not illustrate either the planets in their proper order from the sun, or the relative size of their orbits.

Materials

As chosen by the student.

Procedure

If you make your model very large, such as along a road, you can use a single scale. (This will require parental help and supervision outside school time.) Another way to overcome this problem is to use one scale for the size of the sun, another scale for the size of the planets, and a third scale for the distance to the planets. Whatever scales you use, be sure they are specifically stated.

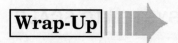

Wrap-Up

1. Which planets group together as the inner planets?

   _____

2. Which of the outer planets is more like an inner planet, and what does it have in common with the inner planets?

   _____

3. How does the size of Jupiter compare to the size of Earth? _____

   _____

4. If these objects were all to the same scale, how would the size of the sun compare with the size of Earth?

   _____

5. Some astronomers have suggested that Pluto was at one time an asteroid or moon. Its status as a planet is in question. As we measure planets in order from Mercury to Neptune, what *generally* happens to the following?

   a. Planetary diameter:

   _____

   b. Average surface temperature:

   _____

   c. Period of revolution:

   _____

6. Ancient people had no telescopes or other ways to magnify celestial objects. Yet, they considered the planets as different from other the objects in the sky. How could you tell a planet from a star by watching it in the night sky without the use of instruments such as a telescope?

   _____

7. The sun has a diameter roughly 100 times the diameter of Earth. What is the volume ratio?

   _____

CHAPTER 27—SKILL SHEET 1: THE MYSTERY OF PLANET X

Instructions

As you read through this Skill Sheet, try to solve the mystery. However, do not say the answer out loud. Raise your hand until the teacher notices you. Do not spoil the fun for others.

The best astronomers, like any good scientists, must also be shrewd detectives. Let us see how careful attention to the clues helps solve a mystery.

In Latin, the accepted style of scientific writing of his time, Galileo wrote the following secret message in his notebook:

The goddess of love imitates the forms of Cynthia.

Who was Cynthia? Did Galileo have a secret lover? What does this have to do with astronomy? If this seems a little strange, stick with it. You will make sense out of Galileo's secret message. In his day, scientists such as Galileo wrote their most important ideas in mystical phrases of Latin text. Such writing was intended for a small group of educated people. (Most people were not educated.)

The English word for planet comes to us from the Greek term, *asteres planetai,* meaning the five special stars that wander among the other stars. (Ancient astronomers saw the planets as points of light and could not see Uranus, Neptune, or Pluto.)

Figure 27-11 shows a portion of the starry night sky over a period of 5 months. You may be able to identify several prominent constellations of the zodiac within this diagram. The zodiac is the region of the sky through which the sun, the moon, and the planets move. These constellations of the zodiac are always high overhead in the tropics. Astronomers prefer to look at negatives of the night sky rather than prints because they can lay one negative on top of another to compare the positions of the stars at different times. Observe the four objects, A, B, C, and D, in each of the five diagrams. One of them is a planet.

1. What is the origin of the word *planeta*?

   _____

2. How can you tell which object in Figure 27-11 is a planet?

   _____

3. Therefore, which of these four objects is the planet?

   _____

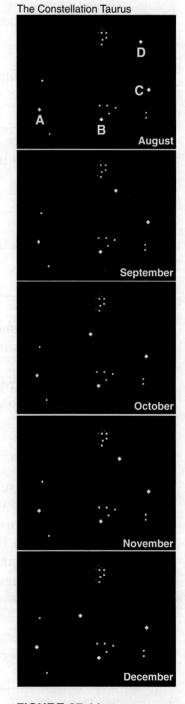

FIGURE 27-11. Star patterns over a period of five months.

Let us call the planet Galileo was describing Planet X. Figure 27-12 shows how it looks as viewed through a telescope or binoculars over a period of several months.

4. Disregarding its changing position in the sky, in what two properties does this object seem to change during the period of observation?

   _____ and _____

This change in apparent shape is called a change in phase.

5. What other familiar celestial object shows phase changes?

   _____

6. When the moon is in the new phase, what does it look like?

   _____

7. When the moon is in the new phase, where is the moon with respect to Earth and the sun?

   _____

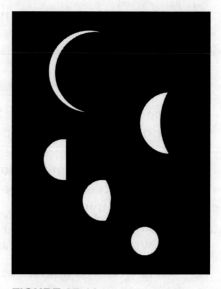

FIGURE 27-12. Five successive photographs of Venus over a several month period.

Therefore, if Planet X shows a new phase, at some point in its orbit, it must be located between Earth and the sun.

8. According to Figure 27-13, which two planets are able to pass between Earth and the sun?

   _____

   Therefore, only these two planets can show a full range of phases (full, gibbous, quarter, crescent, and new, like the moon), as we observe the planets from Earth.
   We also found that our mystery planet shows an obvious change in apparent size (angular diameter).

9. What causes the angular diameter of these celestial objects to change?

   _____

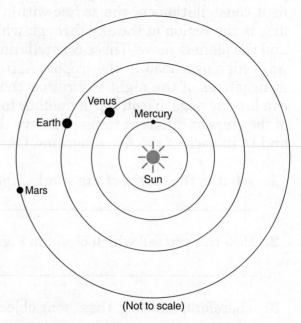

(Not to scale)

FIGURE 27-13. The sun and the four inner (terrestrial) planets.

Copyright © 2007 AMSCO School Publications, Inc.

10. According to Figure 27-13, which of the two innermost planets can come closest to Earth?

_____

11. When it is on the opposite side of the sun, which of these planets can be the farthest from Earth? (Be sure to look at Figure 27-13.)

_____

We have learned that Planet X shows the full range of phases and that it changes greatly in angular diameter as observed from Earth.

12. Planet X must be _____.

13. Who was the Roman goddess of love? _____

14. What celestial object did ancient astronomers call Cynthia, or sometimes Selene?

_____

15. Explain Galileo's secret message "The Goddess of Love Imitates the Forms of Cynthia."

_____

The *Earth Science Reference Tables* has a table of data about the sun, moon, and nine planets. Base your answers to questions 19 through 25 on the information in that table.

16. The largest planet is _____.

17. The smallest planet is _____.

18. Which planet moves at the greatest velocity in its orbit? _____

19. Which planet moves at the slowest velocity in its orbit? _____

20. Why does Pluto move more slowly in its orbit than any other planet?

_____

21. Which planet has the most moons? _____

22. All the planets have night and day, so they must all (a type of motion) _____.

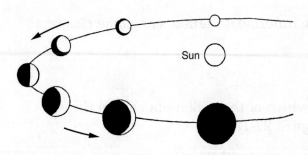

FIGURE 27-14. Venus as viewed from a position in space near Earth.

Figure 27-14 shows Venus as observed from a position in space near Earth. It illustrates why Venus shows the full range of phases and why it changes so much in angular diameter.

Have you ever ridden in a car that is going a little faster than a train moving along an adjacent track? If your car is overtaking the train, from the car it looks as if the train is moving slowly backward. Of course, the train is not really going backward, it just appears to move backward as you pass it in your car, which is moving faster.

Earth moves faster in its orbit than the outer planets. Therefore, when we overtake these planets, it looks as if they are moving backward. This apparent backward motion of the planets is called retrograde motion. Look carefully at the changing position of object D in Figure 27-11.

23. Between which two months does D show retrograde motion (from west to east)?

_____ and _____.

24. What planets are visible in the night sky this evening?

_____

25. If Galileo used a telescope to investigate the night sky, why did his fellow astronomers not use a telescope, also?

_____

Chapter 28
Stars and the Universe

☀ **CHAPTER 28—SKILL SHEET 1: THE ELECTROMAGNETIC SPECTRUM**

Light is one form of electromagnetic energy. All electromagnetic energy travels through space at a speed of 3×10^8 meters per second. That is fast enough to travel around Earth seven times in just one second! Sunlight takes only eight minutes to travel from the sun to Earth.

Figure 28-1 shows that electromagnetic energy includes a wide range of wavelengths. Visible light makes up a small part of the electromagnetic spectrum. However, most of the energy given off by the sun is in the form of visible light. Invisible forms of radiation in sunlight are less intense than visible light.

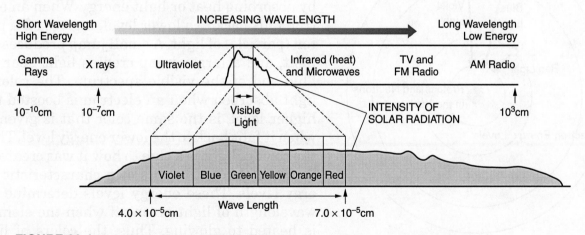

FIGURE 28-1. The sun gives off electromagnetic radiation that is visible and invisible.

White light is a mixture of all the colors of the rainbow. Sir Isaac Newton demonstrated the compound nature of white light when he passed a ray of sunlight through a glass prism as shown in Figure 28-2. Light slows and bends when it travels at an angle through glass. The colors separate because when sunlight enters glass, short wavelengths, such as blue light, are slowed more

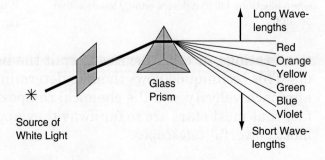

FIGURE 28-2. A glass prism can split light into its component colors.

than long wavelengths, such as red. Therefore the blue end of the spectrum is bent more than the red end, and the colors separate.

1. Name the six colors of the visible spectrum.

_____

2. What part of the electromagnetic spectrum can our eyes detect?

_____

3. Name three forms of electromagnetic energy that our eyes cannot see.

_____

4. Which invisible electromagnetic energy has the shortest wavelength?

_____

5. How can you demonstrate that white light is a mixture of colors?

_____

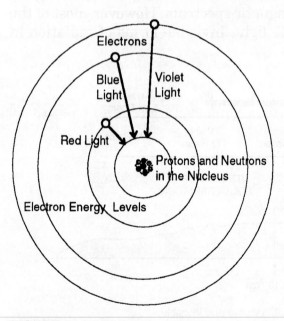

FIGURE 28-3. Various colors of light are given off when electrons fall to different energy levels within an atom.

Electromagnetic energy results from the movement of charged particles. Electrons within the atom can be bumped to a higher energy level by absorbing heat or light energy. When an electron falls back to a lower level, it gives off a photon (particle) of light. A small jump produces red light, and a larger jump creates light near the blue end of the visible spectrum. The color of light absorbed when an electron is boosted to a higher level, is the same color that is given off when it falls back to the lower energy level. Thus, the color of light is a clue to how it was created.

Each element has its own characteristic energy levels. These energy levels determine the wavelength of light given off when the element is heated to glowing. Thus, the colors of light given off by an element are like its fingerprints. Each element can be identified by the colors in its own spectrum.

Astronomers can use glass prisms, diffraction gratings, or other devices to split the light from distant stars into a spectrum of colors. This technique allows them to determine the mass, surface temperature, recession or approach velocity, and the chemical composition of stars. They can do this in spite of the fact that most stars are so far away they appear as tiny points of light even in the world's most powerful telescopes.

6. What happens to electrons that absorb light energy?

_____

7. What change within an atom causes it to give off light?

_____

8. What determines the color of the light given off?

_____

9. What color of visible light is produced by the longest wavelength of visible light?

_____

10. What color of visible light is produced by the shortest wavelength of visible light?

_____

For the next three questions, use the Electromagnetic Spectrum table (page 14) from the *Earth Science Reference Tables.*

11. What is the average wavelength of X-rays?_____

12. What is the range of wavelengths of visible light?

_____

13. Which has a longer wavelength, ultraviolet rays or infrared rays?

_____

My Notes

CHAPTER 28—LAB 1: ELECTROMAGNETIC SPECTRA

Introduction

There are four types of electromagnetic spectra. The **continuous spectrum** shows a single band of the full range of visible colors; red, orange, yellow, green, blue, and violet. A continuous spectrum is usually produced by a hot source that emits radiation throughout the visible region of wavelengths.

A **filtered** spectrum is like a continuous spectrum, but much brighter in a certain part. This can be at either the red end, the blue end of the spectrum, or at any color in the spectrum.

A **bright line** spectrum is discontinuous. It contains narrow bands of color separated by wider dark areas. The glowing gas of a single element often creates it. A bright line spectrum is produced by electron jumps of several distinct sizes.

The **dark line** spectrum is a continuous spectrum that contains a few dark bands. It is produced when white light passes through a cloud of cool gas. The colors absorbed by the gas are the same colors that would be produced as bright lines if the same gas were heated enough to glow.

Objective

To investigate electromagnetic spectra.

Materials

Diffraction spectroscopes, gas tubes and lightbulb, red and blue filters, wall chart of spectra

Procedure

Your teacher will show you how to use the spectroscope. For each light source, record the name of the source, the colors that are visible in the spectrum, and the type of spectrum (continuous, filtered, bright line, or dark line).

TABLE 28-2. Spectrum Lab Data

Number	Spectrum source	Colors present	Spectrum type
1			
2			
3			
4			
5			
6			
7			

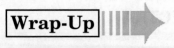
Wrap-Up

1. Visible light is one kind of _____ radiation.

2. What color of light has a wavelength barely longer than green light?
   _____

3. What is created when white light is split into its colors?
   _____

4. How is a continuous spectrum changed into a filtered spectrum?
   _____

5. How can scientists use light to identify an element?
   _____

6. What is light? _____

7. Why is white not really a color? _____
   _____

8. How is light produced on the surface of the sun?
   _____
   _____

9. How does this activity illustrate the quantum nature of energy?
   _____
   _____

10. Electromagnetic energy, such as visible light, has a "split personality." Some
    experiments prove it is one thing, while other experiments prove it is something
    very different. State the two "personalities" of electromagnetic energy and explain
    an experiment that proves each.
    _____
    _____
    _____

CHAPTER 28—LAB 2: IS THE SUN AN AVERAGE STAR?

Introduction

Clearly, the easiest star to study is our sun. However, can we reasonably assume that most stars are like the sun? On the other hand, is the sun unusual in more than how close it is to Earth? To answer this question, we will use a graph developed by two astronomers. The graph is called the Hertzsprung-Russell (H-R) diagram.

Objective

You will plot the characteristics of the sun and other stars on the Hertzsprung-Russell diagram, to determine whether the sun is a typical star. The Hertzsprung-Russell diagram will show whether the sun is much brighter, dimmer, hotter, or cooler than most other stars. If that is what you find, you should be careful in thinking of our sun as a typical star. If, on the other hand, the sun is near the center of the group of stars, it would appear to be typical.

However, in this lab you will find that there is another important issue in using this method. You must also consider the group of stars that you use to compare with the sun. If you select an atypical group, the comparison will not be valid. How can you select the most representative group of stars?

Materials

Hertzsprung-Russell (H-R) diagram

Procedure

Part I

Table 28-2 lists the twenty brightest stars we see from Earth. If you have observed the night sky, these star and constellation names may already be familiar to you. Plot the sun as a letter *S* on the H-R diagram (Figure 28-4), and then plot each of the other stars as a small *x*.

TABLE 28-2. The 20 Brightest Stars

Name	Constellation	Visual Magnitude	Spectral Type	Absolute Magnitude
Sun	None	−26.72	G2	+4.8
Sirius	Canis Major	−1.46	A1	+1.4
Canopus	Carina	−0.72	F0	−3.1
Rigil	Centaurus	−0.01	G2	+4.4
Arcturus	Boötes	−0.06	K2	−0.3
Vega	Lyra	0.04	A0	+0.5
Capella	Auriga	0.05	G8	−0.67
Rigel	Orion	0.14	B8	−7.1
Procyon	Canis Minor	0.37	F5	+2.6
Betelgeuse	Orion	0.41	M2	−5.6
Achernar	Eridanus	0.51	B3	−2.3
Hadar	Centaurus	0.63	B1	−5.2
Altair	Aquilla	0.76	A7	+2.2
Aldebaran	Taurus	0.86	K5	−0.7
Acrux	Southern Cross	0.90	B2	−3.5
Spica	Virgo	0.91	B1	−3.3
Antares	Scorpius	0.92	M1	+5.1
Fomalhaut	Pices Austrinis	1.15	A3	2.0
Pollux	Gemini	1.16	K0	+1.0
Deneb	Cygnus	1.26	A2	−7.1

1. How does the sun compare with the other bright stars we see in the sky?

_____

2. Would you call the sun a "typical" member of this group of bright stars? _____

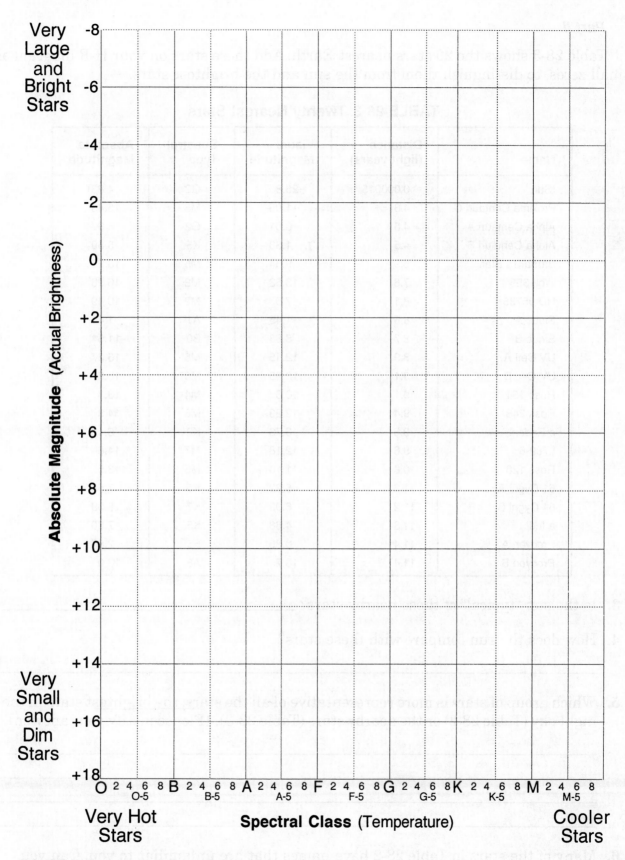

FIGURE 28-4. Hertzsprung-Russell diagram.

Part II

Table 28-3 shows the 20 stars nearest Earth. Add these stars on your H-R diagram as small zeros, to distinguish them from the sun and the brightest stars.

TABLE 28-3. Twenty Nearest Stars

Name	Distance (light years)	Visual Magnitude	Spectral Type	Absolute Magnitude
(Sun	0.000015	−26.8	G2	4.83)
Proxima Centauri C	4.3	11.05	M5	15.45
Alpha Centauri A	4.5	−0.01	G2	4.3
Alpha Centauri B	4.5	1.33	K5	5.69
Barnard's Star	5.9	9.54	M5	13.25
Wolf 359	7.6	13.53	M8	16.68
HD 95735	8.1	7.5	M2	10.49
Sirius A	8.6	−1.45	A1	1.41
Sirius B	8.7	8.68	B0	11.54
UV Ceti A	8.9	12.45	M5	15.27
UV Ceti B	9.1	12.95	M6	15.8
Ross 154	9.1	10.6	M4	13.3
Ross 248	9.4	12.29	M6	14.8
e Eridani	9.5	3.73	K2	6.13
L789-6	9.6	12.18	M7	14.6
Ross 128	10.8	11.10	M5	13.5
61 Cygni A	11.2	5.22	K5	7.58
61 Cygni B	11.2	6.03	K7	8.39
e Indi	11.3	4.68	K5	7.00
Procyon A	11.4	0.35	F5	2.65
Procyon B	11.4	10.7	A8	13.0

3. Is the sun "typical" of these nearby stars? _____

4. How does the sun compare with these stars?

_____

5. Which group of stars is more representative of all the stars, the brightest stars in the night sky (Table 28-2) or the nearby stars (Table 28-3)? (Please justify your answer.)

_____

_____

_____

6. Many of the stars in Table 28-2 have names that are unfamiliar to you. Can you suggest why the stars in Table 28-2 have such unusual names?

_____

7. Use the Luminosity and Temperature diagram in the *Earth Science Reference Tables* to write star groups on your H-R graph.

8. What group of stars seems to be most common?

_____

Wrap-Up ▐▐▐▶

1. What two variables are graphed on an H-R diagram?

_____

2. What is the surface temperature of the sun? _____

3. What color stars are the hottest stars we see? _____
   What color stars are the coolest stars we see? _____

4. What is the general relationship between star mass and energy output?

_____

5. How does the selection of stars for comparison affect our judgment of whether the sun is truly a typical star?

_____

_____

6. The selection of a small number of things to represent a larger group is a common practice in our lives. Where else is this kind of sampling used?

_____

_____

7. What are the Nielsen Ratings? _____

_____

8. If you were to do your own TV popularity rating, how might you select households to monitor that would be most like Table 28-2?

_____

_____

9. How might you select households to monitor that would be most like Table 28-3?

_____

10. If we consider the sun an average star, how can we justify that evaluation? Base your answer to this question on the results of this investigation. (Please explain your answer in one or more complete sentences.)

☀ CHAPTER 28—SKILL SHEET 2: DYNAMICS OF STARS

Introduction

When you think of certainty in your life you may think of the march of the sun across the sky each day. When the sun sinks below the horizon, causing night, you are confident that it will rise again at a precisely predictable time and place a few hours later. If the sun were to change significantly, the effect on planet Earth could be catastrophic. In fact, the sun will change as it completes its life cycle.

Find photographs or Internet images that illustrate the various stages in stellar evolution and paste them on Figure 28-5, from the *Earth Science Reference Tables*. Label each image with the name of a star that is currently in that stage of stellar evolution. On your diagram, draw a line or lines to show the evolution of stars from their formation to the time they no longer give off significant amounts of visible electromagnetic energy.

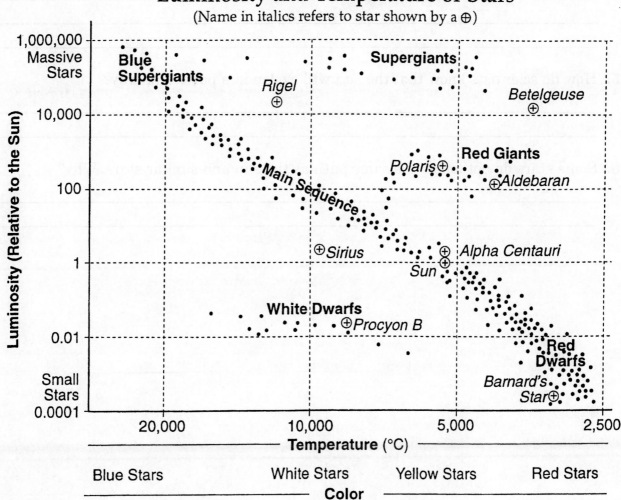

FIGURE 28-5. The Luminosity and Temperature of Stars chart, a form of the Hertzsprung-Russell diagram, is a useful tool to trace the evolution of stars.

Wrap-Up

1. How long will it take for the sun to complete its evolution?

   _____

2. How do stars make electromagnetic energy? _____

   _____

   _____

3. Why do stars change? _____

   _____

4. Do you think that future changes in the sun will play an important part in the future of humans? Please explain your answer.

   _____

   _____

   _____

5. How do scientists know that the sun will evolve in a particular way?

   _____

   _____

6. Some stars do not follow the same path as the sun and similar stars. Why?

   _____

   _____

Appendix

SKILL SHEET 1: AT YOUR FINGERTIPS

Without question *The Earth Science Reference Tables* is your most important resource for the Earth science Regents exam. Knowledge of the information in the ESRT and how to use it will be of critical importance in your exam performance. Please keep in mind that completing this Skill Sheet should be a learning activity. If you are unable to answer any of these items, your teacher can help you use this an opportunity to pick up new knowledge.

Reference Table Page 1

1. To the nearest 0.1 cm, how wide is this sheet of paper? _____

   How wide is it in meters? _____

2. Which of the radioactive substances listed on this page has the shortest half-life?

   _____

3. Write the half-life of uranium-238 as a standard number.

   _____

4. If you started with 100 grams of K-40, how much K-40 would remain after 3.9×10^9 years?

   _____

5. What common substance requires the most energy to heat a unit mass by 10°C?

   _____

6. Which requires more energy, melting a 10-gram ice cube or evaporating 10 g of water?

   _____

7. If a student estimated the volume of a rock to be 20 cm³, but careful measurement showed its true volume was 25 cm³, what is the percent deviation of the estimate? (Please show your work.)

8. In Figure A-1, what is the average gradient from point A to point B?

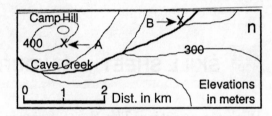

FIGURE A-1. Camp Hill map.

Reference Table Pages 2 and 3

9. In what landscape region do you live?

_____

10. What is the landscape region around Old Forge, New York?

_____

11. The Catskill Mountains are a part of what larger landscape area?

_____

12. If you want to drive from Albany to Buffalo, in what direction must you travel?

_____

13. What are the approximate terrestrial coordinates of Watertown, New York?

_____

14. What is the numerical age of the bedrock around Syracuse, New York?

_____

15. What is the metric distance from Syracuse to Utica, New York?

_____

16. As water flows from Lake Erie into Lake Ontario, most of the change in elevation is at Niagara Falls. What is the total change in elevation?

_____

17. Which kind of rock is most common in the Catskills? _____

18. What New York landscape region has the oldest bedrock?

_____

Reference Table Page 4

19. What ocean current keeps Europe relatively warm?

_____

20. How do local ocean currents affect the climate along the western coast of South America?

_____

21. What major ocean current can be found at 50°S, 50°W?

_____

Reference Table Page 5

22. With respect to Africa, in what direction is South America drifting? _____

23. What kind of plate boundary is the Mid-Atlantic Ridge?

_____

24. What is the major active fault in the Western United States?

_____

25. What has caused the growth of the Himalayan Mountains, north of India?

_____

26. Where is the only major mantle hot spot/plume in the continental United States?

_____

Reference Table Page 6

27. What is the final step in the formation of sediment? _____

28. What name is given to sediment composed of particles 1 cm across? _____

29. How does gabbro differ from basalt? _____

_____

30. What is the most abundant mineral in diorite?

_____

31. How fast must the current in a stream be to transport the smallest boulders?

_____

32. What five minerals are common in basalt? _____

_____

Reference Table Page 7

33. What minerals are most common in sandstone?

_____

34. Which clastic (fragmental) rock is composed of the smallest particles?

_____

35. What mineral is most abundant in rock salt? (The mineral is not called salt.)

_____

36. What mineral family is common in slate, phyllite, schist, and gneiss?

_____

37. What is the texture of quartzite? _____

38. What metamorphic rock is composed primarily of calcite? _____

Reference Table Pages 8 and 9

39. When did North America split from Africa and Europe?

_____

40. How old is Earth? _____

41. What is the first period of the Paleozoic Era? _____

42. What two periods are not represented in the bedrock of New York State?

_____ and _____

43. What animal group first evolved about the same time as the dinosaurs?

_____

Reference Table Page 10

44. At what two depths within Earth is the temperature above the melting point of rock? _____ and _____

45. What is Earth's radius in kilometers? _____

What is Earth's diameter in kilometers? _____

46. What is the composition of Earth's core? _____

47. Which layer of Earth is the least dense? _____

Which layer is most dense? _____

48. In what part of Earth does the temperature increase fastest with depth?

_____

Reference Table Page 11

49. What are the two most common elements in the oceans?

_____ and _____

50. What element makes up about 6% of crustal rocks by mass, but only about 0.5% by volume?

_____

51. How long does it take an S-wave to travel 6000 km? _____

How long does it take a P-wave to travel 6000 km? _____

52. How far away is the epicenter if the P-wave arrives 5 minutes before the S-wave?

_____

53. How far can a P-wave travel in 5 minutes, 40 seconds? _____

Reference Table Page 12

54. If the wet bulb reads 4°C and the dry bulb records 12°C, what is the dew point?

_____

55. What is the relative humidity in the conditions specified in question 54? _____

56. What does a negative dew-point temperature indicate?

_____

Reference Table Page 13

57. What Kelvin temperature is equivalent to 0°C? _____

What is this temperature in Fahrenheit? _____

58. Express normal atmospheric pressure in inches of mercury.

_____

Base your answers to questions 59–61 on Figure A-2.

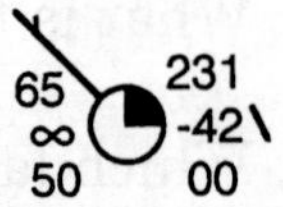

59. What is the Fahrenheit temperature at this weather station?

_____

What is the Celsius temperature? _____

```
        231
65  ⟋⟍  -42\
∞ ( )
50      00
```

FIGURE A-2. Weather station model

60. What is the atmospheric pressure? _____

Is it rising or falling? _____

61. What is the wind speed? _____

What is the wind direction? _____

What is the cloud cover? _____

62. A maritime tropical air mass is _____ and _____.

(You should know this without using the *Reference Tables*.)

Reference Table Page 14

63. How does the air temperature change as you go higher within the mesosphere?

_____

64. What name has been given to the boundary between the lowest two layers of the atmosphere?

_____

65. What form of electromagnetic energy has a wavelength just beyond our visual range?

_____

66. What is the wavelength range of visible light? _____

67. What is the prevailing wind direction 45° south of the equator?

_____

68. Why is precipitation abundant near the equator? _____

_____

Reference Table Page 15

69. Although Barnard's Star is relatively close to us, it is hard to see. Compared to the sun, how much light does Barnard's Star give off?

_____

70. The North Star is similar to the sun in _____ and _____.

How is it very different? _____

71. Rigel and Betelgeuse are two of the brightest stars in the winter constellation Orion. How does Betelgeuse appear different from Rigel?

_____

72. Of the nine planets, which one rotates fastest on its axis? _____

73. Which planet has the most eccentric (flattened) orbit? _____

74. Which planet is closest in size to Earth? _____

75. Approximately how much larger is the sun's diameter than the diameter of Earth?

_____

Reference Table Page 16

76. What mineral has a nonmetallic luster, scratches glass, and is often pink?

_____

77. Name two common minerals that are composed of just one chemical element.

_____ and _____

78. What is the most obvious difference between amphibole and pyroxene?

_____

79. What other minerals form crystals similar in shape to pyrite?

_____

My Notes